Nature's Ghosts

SOPHIE YEO

Nature's Ghosts

The world we lost and how to bring it back

Harper North

HarperNorth
Windmill Green
24 Mount Street
Manchester M2 3NX

A division of
HarperCollins*Publishers*
1 London Bridge Street
London SE1 9GF

www.harpercollins.co.uk

HarperCollins*Publishers*
Macken House, 39/40 Mayor Street Upper
Dublin 1, D01 C9W8

First published by HarperNorth in 2024

3 5 7 9 10 8 6 4 2

A catalogue record for this book
is available from the British Library

HB ISBN: 978-0-00-847412-6

Printed and bound in the UK using 100%
renewable electricity at CPI Group (UK) Ltd

For Jack

Contents

PROLOGUE

The Deserted Village

A few years ago, I developed an obsession with seeking out lost features of the countryside. Old roads, sunken through centuries of footfall. Ancient trees that once marked the boundaries between fields that had since been combined. Standing stones erected thousands of years ago for reasons unknown.

It was this peculiar hobby that took me, one springtime day, to a field in Northumberland, an hour or so away from my home. My curiosity had been piqued when, looking over the Ordnance Survey map the previous evening, I saw 'South Middleton Village' printed in gothic script. It stood out against the business-like fonts of the settlements that surrounded it, and so I did some research. South Middleton, it turned out, was a deserted medieval village. I could find almost nothing about it online, other than that it first appeared in documents in the thirteenth century, suffered a declining population through the following years, and was fully abandoned by 1762. That did not seem so long ago: I had eaten lunch in pubs older than that.

So I set off the next day to see it in real life. I don't know what I expected to find, exactly. Perhaps some tumbledown buildings, or at the very least some rubble – something that might attest to the centuries of life that had played out here. Arriving at my destination, however, it became apparent that 'deserted medieval village' was something of an understatement. South Middleton was no Pompeii, no Californian gold rush ghost town. Looking over the field at where the village had once been, I saw ... nothing. There were no ruins, no foundations. Just rough grass and a line of leafless trees. Where once there had been buildings, gardens, community, there was now a working farm. I heard a collie dog bark in the distance. If I hadn't seen its name on the map, I would not have known there had ever been a village here at all.

I entered the field and tried to resurrect it in my mind's eye. From my reading, I knew there had once been two lines of houses, facing onto a rectangular village green. Behind the houses were small gardens, and, beyond that, the communal farmland where the villagers would have grown their crops. I wanted to see the faces of those people, to empathise with the family who said the final goodbye to this place some three centuries ago – but it was difficult. Time had not left me much to go on.

However, as I walked up and down the field, South Middleton slowly began to rise from the dead. Not figuratively. Literally. The houses and the gardens had not entirely vanished. They had left their mark in raised piles of earth: gentle undulations that seemed natural but weren't. They bulged out of the pasture, tracing the contours of the houses, gardens and green. My eyes gradually tuned into the patterns, and I felt the rising of the soil underfoot.

Suddenly, I was no longer in a field. I was in a timber longhouse, sitting with a family as they waited for the snow to subside, huddling among the livestock with which they shared their space. I crouched alongside a woman as she parted the mud to plant herbs in her plot, relishing the scent of rosemary and sage that wafted from the leaves crushed lightly in her hands. I ran alongside the children as they laughed, played and chased one another across the village green. I felt the weariness of the final hold-outs, abandoned by their friends and neighbours, and wondered what had finally pushed them, too, to leave their home. Was it loneliness? A bad harvest? A sickness; a death? In that moment, South Middleton seemed so much more than an oddity on a map. There had been life here. Now it was gone.

Of course, there had been a time before South Middleton even existed. It was unlikely that the villagers were the first people to know this acre of England. To get here, I had walked along an old Roman road that had passed through an even older Iron Age fort, perched upon a crag. There, I saw a kestrel flying across the field below and sat to watch it for a while, struck by the novelty of having a bird's-eye view of a bird.

The view stretched for miles. It occurred to me how different the scenery must have looked to the occupants of this fort, thousands of years ago. Would a kestrel have been anything remarkable for them, or would it have taken nothing less than an eagle to steal their attention? Might they have watched aurochs roaming the land down below? Would the horizon have been cloaked by forest or farm? Earlier that day, I had passed a burial mound that was even older than the fort, constructed sometime during the Late Neolithic or Early Bronze Age, at a time when the environment would have looked totally different again. How far had the wildwood retreated, I wondered, by the time that corpse was sent to the grave? Would theirs have been a world of darkness or light?

So many layers. So many landscapes. Not to mention the moments that have not been immortalized in relics, either because the human presence was more fleeting – a quiet afternoon spent picking hazelnuts, perhaps, or a silent prayer offered to some unknown god – or because there were no humans around at all. I thought too of the non-human life that might have cycled through the field where South Middleton had once stood: the plants and animals that had shaped this place through the last thousands and millions of years. Had a mammoth ever stridden through the boundaries of that future herb garden? Had there ever been a litter of wolf pups to foreshadow those laughing children on the village green? There was no way to tell – at least, not on this occasion.

The landscape is full of ghosts. Tune into the minutiae of the earth, and you will start to feel their presence. Some of those details can be perceived with the naked eye: the undulations of South Middleton, for instance, or the ditches of an Iron Age hillfort. They are the land's archives of the people who passed through it. The ghosts that haunt such places tend to be those of the humans that occupied them.

But the ecosystems that surrounded such places left their imprint on the landscape, too. It is these ghosts that are the focus of this book. At some point, my obsession with seeking out remnants of human history turned into an obsession with peeling back the layers of natural history: of understanding how and why the living world has changed through time. The evidence for this transformation is harder to see, but it is there. To unveil it requires deep ecological knowledge, special equipment, and access to obscure written documents and museum collections. Throughout my reporting, I have read about, spoken to and kept company with people who possessed all three.

My motivation for understanding the history of the environment is that I believe it can and should inform how nature conservation happens today. Those who invoke the past are often dismissed when it comes to restoration: the subject is seen as unfashionable, with those who talk of it accused of prioritising romanticism or nostalgia over realism and pragmatism. This has particularly been the case with rewilding, where practitioners are often keen to stress that they are not seeking to return the world to some golden moment in time, but rather to unleash nature from its modern shackles and set it free upon a twenty-first century path.

Because of this, the conversation has become too binary. Conservation is treated as a choice between backwards and forwards, past and future, re-creation and innovation. In each case, it is usually the latter that is seen as the safer and more respectable option.

I may as well say now that I do not believe that we can, or should, attempt to rewind the world to a specific point in time, for reasons that will become apparent. But that does not mean that we should treat the past as irrelevant. Because it is equally clear to me that the ecosystems of ten thousand, one thousand, or even one hundred years ago were wilder, richer and more enchanting than that which passes as nature today. At a time of unprecedented ecological collapse, it is our duty to restore something of that lost world. By peering into the annals of the earth, we can learn how to begin.

ONE

When Humans Were Wild

Once, the world was wild. Plants grew where birds dropped their seeds. Those seeds grew into flowers, shrubs and trees, whose fruits and foliage fed the herbivores that roamed the landscape. Those herbivores travelled with the seasons, crossing undrawn borders as they migrated through the unnamed continents. Rivers, too, carved out their own routes, braiding through rock and soil, spilling outwards into floodplains and wetlands. Fires caught where lightning struck the ground or when the summer burned too hot. The climate warmed and cooled in long and natural cycles, causing glaciers to expand and shrink, and sea levels to rise and fall.

For a while, humans were just another part of that wilderness: creatures no more separate from their surroundings than any other. The history of hominins is long and complicated; a family tree branching off in several directions, rather than an easy evolutionary flow from one species into the next. In short, the human lineage split from chimpanzees around 7 million years ago, with the *Homo* genus – which includes us, but also

Neanderthals, for instance – arriving on the scene just over 4 million years later. Our own species, *Homo sapiens*, is still young, relatively speaking, appearing somewhere between 200,000 and 300,000 years ago.

The lives of our ancient relatives are still poorly understood, but of one thing we can be sure: like the plants and the animals, they too followed the earth's natural rhythms. The impact of an early human hunting for meat was no more significant than a lion picking off its prey. Their foraging made no more of a dent in the ecosystem than the flocks of birds that feasted upon the same berries. The journey of a seed stuck to the leg of a human traveller was not materially different to that of one caught in the fur of a mammoth. The fact of being human did not make our forebears any less a part of nature than the beasts with which they shared the earth.

But the once-wild world has now been tamed – by us. The forests, steppes and wetlands of prehistory have been transformed into cities, fields and pastureland. The wild beasts that our ancestors hunted have either been domesticated or confined behind fences; those that still roam 'free' have had their migratory routes curtailed by tarmac and concrete. The ocean has been emptied of fish. Insects end their lives on the windscreens of cars. Fires are started by the flick of a cigarette. The climate is unnaturally hot, thanks to more than a century of heavy fossil fuel combustion.

At some point, on the journey from tree-climber to toolmaker to city-dweller, our relationship with nature shifted. We switched from cogs in the wilderness to the masters of our environment. When did that transition take place?

In some ways, humans have been transcending the limitations of their bodies since the very start of their story. The first stone tools predate the appearance of the *Homo* genus by 500,000 years, suggesting they were made by a more primitive branch of the hominin tribe. Growing more sophisticated with time, these tools changed how humans engaged with the world, transforming us from scavengers to hunters. Tool use alone, however, does not set humans apart from other wild animals: gorillas, elephants and even sea otters are among the many other creatures that also possess this skill.

Fire was another crucial turning point in the development of our species. From around 400,000 years ago, there is evidence that hearths were kept burning in caves across Europe, the Middle East, Africa and Asia. These provided warmth and deterred predators and, crucially, allowed those who tended them to cook their butchered meat.[1] But nor are humans unique in their control of flames. Aboriginal Australians, for instance, have long observed that raptors will transport flaming sticks from bushfires and drop them into unburned savannahs to flush out their prey.

Together, though, toolmaking and fire-starting unlocked something unique. They allowed humans to sculpt the landscape around them, re-shaping the wilderness to meet their needs. It was this skill – this urge to transform – that ultimately set our species apart from beasts. No other creature has looked out upon the world around them and thought: I can do better.

Intellectually speaking, our forebears would have been up for the challenge of landscape transformation from at least 100,000 years ago. The skulls of humans living at this time were essentially the same as those of present-day humans. 'Assuming the brain was as modern as the box that held it, our African ancestors theoretically could have discovered relativity, built space telescopes, written novels and love songs,' writes the palaeontologist Nicholas R. Longrich. 'Their bones say they were just as human as we are.'[2]

Our ancestors, of course, did not invent relativity or build telescopes. But the moment at which they began to transform the wilderness into their personal garden is harder to identify. Attempts to do so have been held back by misunderstandings of deep time and, indeed, the failure of modern humans to recognize the brilliance of those who came before them. It is only in the past few decades that scientists have really started to acknowledge just how profoundly, and for how long, humans have been shaping the face of the earth.

Deep time is a relatively recent discovery. The received wisdom in Europe, until the eighteenth century, was that God had created the earth around 6,000 years ago. On 23 October 4004 BCE, to be precise. It was the Scottish geologist, James Hutton, who suggested in 1788 that the world was far older – the product not of divine intervention, but of cycles of erosion and uplift across millions of years, which had created the stratified patterns that could still be seen in the exposed surfaces of cliffs and gullies. Even the very oldest rocks, he wrote, were 'furnished from the ruins of former continents'.

Racist opinions, formed on the back of colonial expansion, meant that many refused to countenance the idea that prehistoric people could have been capable architects of their environment. The general assumption was that ancient humans would have behaved similarly to the Indigenous people encountered on foreign shores, who were regarded as little more than wild animals. Charles Darwin, writing in *The Descent of Man* in 1871, concluded that 'there can hardly be a doubt that we are descended from barbarians', adding that he would rather trace his origins back to a monkey or a baboon than the 'savage' inhabitants of Tierra del Fuego whom he had witnessed on one of his voyages. The antiquarian John Aubrey, writing in 1659, assumed that the landscape around Stonehenge must have been a 'shady dismal wood' whose inhabitants were 'two or three degrees I suppose less savage than the Americans'.[3]

Even today, conservationists often still draw the baseline for what is considered 'natural' just a few centuries in the past, despite the fact that the historic impacts of Indigenous people on the landscape are now well understood. In America, the arrival of Christopher Columbus in 1492 is often considered to have been the moment when humans began making their mark on what had, until that point, been pristine wilderness – a division that ignores the evidence for cultivation, burning, building, hunting and deforestation undertaken by the Native Americans who had already lived on the continent for thousands of years.

Those who take a deeper view of time tend to regard the invention of agriculture as the dividing line between the 'natural' and anthropogenic environment. Until this moment, humans had been hunter-gatherers: their sustenance came from the plants and animals that they plucked and speared in the wild. But around 12,000 years ago, humans realised there was an easier way to satisfy their needs. By domesticating plants and animals, they could ensure – in theory – a plentiful supply of food, in convenient locations, that they could store and consume at their leisure. From its origins in the Near East, agriculture spread across Europe and finally into Britain around 4,000 BCE. This was the start of the period known as the Neolithic.

Farming changed the face of the landscape. Finally, humans were able to exercise extensive control over what plants grew and where. It is easy to see why this transition is regarded as the moment when nature became artificial: over time, self-willed ecosystems were replaced by weedy fields and wild animals pushed out by livestock. These changes were designed to benefit a small subset of the planet's inhabitants to the detriment of the natural world at large. The arrival of agriculture marked the beginning of a new way of life, leading to settled communities and the ability to raise more children, and ultimately to the rise of the pathways and villages required to connect and house the burgeoning population. Compared to the fierce and thorny world of the hunter-gatherers, the landscape must have appeared tame and navigable – perhaps even safe.

But those hunter-gatherers were not quite as passive as it is often thought. In fact, their influence on the landscape has been seriously underestimated. In 2021, a group of academics, comprising archaeologists, ecologists and anthropologists, published a paper that challenged the notion that the transformation of nature has been 'mostly recent and inherently destructive'. By reconstructing long-term patterns of population and land-use, they revealed that, by 10,000 BCE, nearly three-quarters of the earth's land surface had been occupied and shaped by humans, and that less than 4 per cent of temperate woodlands remained undisturbed. Given that agriculture had barely begun by this point, the architects of this change could not have been farmers.[4]

The last Ice Age ended around 11,700 years ago. This marked the beginning of the Holocene, a period of warm and stable temperatures that continues to this day.[5] For the first few millennia of this epoch, hunting and gathering remained the only way of collecting food. This period, when Europe was green and temperate – but not yet cultivated – is known as the Mesolithic. Certainly, the hunter-gatherers of this time were not as aggressive in shaping the environment as the farmers who succeeded them. It remains difficult to find unambiguous traces of their presence within the archaeological record, demonstrating that they mostly trod lightly upon the earth. But that does not mean they left no footprints at all.

Throughout the Mesolithic, humans became increasingly adept at manipulating their environment: they knew what they wanted and how to get it. By opening glades within the forest, most often through burning, hunter-gatherers created the conditions for a natural larder to emerge. Following the obliteration of the trees, edible plants – including hazelnuts, acorns, fruit, berries and bracken – would have sprouted from the sun-warmed soil.[6] The young shoots of the regenerating trees would have attracted herbivores, too – aurochs, deer, wild boar and elk – putting them within easy reach of the hunter's spear: the meat to go with the greens.[7] Evidence suggests that this kind of disturbance was surprisingly common, causing long-term changes to the structure of the forest: the wildwood was not quite as wild as it may have seemed.

Such tactics represented the halfway point along the road to farming. The plants and animals attracted into these glades were not sown or reared, and they had not yet been domesticated: they remained, essentially, wild. Nonetheless, these proto-supermarkets reduced the uncertainty of where the next meal might come from, and saved hunters from having to drag a bloody carcass across a predator-filled landscape in order to feed their families. Archaeologists have even speculated whether these burned-out patches perhaps led to the very first notions of land ownership.[8] Would the hunter who started the fire have enjoyed special rights to the produce of his glade? It is interesting to contemplate just how quickly humans might have seized upon the concept that the quickest way to claim the land is to transform it for one's own benefit.

It has also been suggested that hunter-gatherers permanently altered the behaviour of the herbivores they stalked. By effectively herding a lot of animals into a smaller area, the hunters gave themselves the luxury of choice. Roe deer bones recovered from a Mesolithic summer camp in Germany suggest that hunters overwhelmingly targeted adult females and their babies over the bulkier males – a preference that may account for the species' unusually high reproduction rate to this day.[9] Beavers are another animal that alter their reproductive behaviour in response to hunting pressure. By deliberately targeting the young, females can be 'tricked' into producing more offspring the following year to compensate for the loss, a technique used by Native Americans in the nineteenth century to secure a steady supply of beaver pelts without depleting the population as a whole.[10]

Mesolithic people also contributed to the spread of invasive species, intentionally rearranging the jigsaw pieces of the natural world because it suited them. Wild boar, for instance, is often thought to be native to Ireland when, in fact, it is more likely to have been introduced by early settlers around 9,000 years ago. Transporting an aggressive hairy pig across the Irish Sea in a cramped boat was surely not a pleasant experience – it is possible they brought over hand-reared babies to reduce the ordeal – but the journey was quick and the payoff would have been huge. The extreme swings of temperatures during the last Ice Age had effectively cleared Britain and Ireland of its large herbivores. Once conditions had stabilised, the extirpated animals could return to England across the land bridge that still connected it to Europe. Ireland, however, had already lost its link to the continent, and hosted a slim array of fauna as a result. If hunter-gatherers were to survive there, they would need something to eat – and the meaty boar made an excellent meal.[11]

The advantages of the boar weren't only culinary. Ireland would have been forested at this point, making it challenging for humans to navigate. Wild boar act as natural bulldozers, overturning the soil to seek out the roots, tubers and insects below the surface. This activity would have helped to create clearings in the woodland, allowing the new occupants to make inroads from the estuaries and river valleys into the wild heart of the country. Nor would humans have been the only species to benefit from their rooting. Flowers, birds and butterflies would have found food and habitat among the turned-up mud. For ground-nesting birds, on the other hand, whose nests and eggs are vulnerable to predation by the pigs, their presence may not have been so welcome.[12]

Wild boar may not have been the only creatures that hunter-gatherers brought across the sea. Lynx and wildcats also appeared in Ireland for the first time during the Mesolithic, with archaeologists speculating that they may have been introduced for their luxurious fur. The reappearance of brown bears during this period is harder to explain; it is not clear that the presence of the predator would have had any upsides for the human settlers. It is, however, just about possible that the species was transported to the island for spiritual reasons; elsewhere, bear bones have been found stained with ochre, suggesting they played a role in religious ceremonies.[13]

Indeed, a group of archaeologists recently suggested that the prevailing assumptions about Mesolithic landscape change had been overly practical, emphasising the procurance of food and materials to the exclusion of anything else.[14] This approach frames the hunter-gatherers as little more than 'automatons', the archaeologists wrote, and ignores the rising body of evidence pointing to their rich spiritual and cultural lives. These were people who buried their dead and wore deer-skull headdresses, possibly for shamanic purposes. They may well have feared the unmapped forests beyond the known world of their camps. Seen in this context, it seems entirely plausible that our ancestors might have thought of more than food as they blazed their way through the wilderness. Perhaps the glades they created were not only repositories of meat and fruit, but also havens of light and safety within a dark and dangerous world.

Frightening as the forests of the Mesolithic may have been, they were nothing compared to the frozen landscapes of the epoch that preceded it. The Pleistocene began 2.6 million years ago. For millennia, walls of ice advanced and retreated across the face of the earth, like a slow-moving tide, in a cycle of periods known as glacials and interglacials. The last glacial cycle began around 115,000 years ago, reaching its peak around 20,000 years ago. This harsh environment was the backdrop to our species' first forays across Europe.

The landscape might have seemed bare, but it certainly wasn't barren. At their greatest extent, the glaciers covered around a third of the planet's land surface. Beyond their edges lay a varied world of scattered trees, tundra, bog and steppe, supporting megafauna including mammoth, woolly rhinoceros, giant deer, cave bear and sabre-toothed cat. These are the creatures that would have haunted the dreams of the first Europeans. They certainly featured in their artwork, sketched onto cave walls in ochre and charcoal by flickering firelight.

For most of the Pleistocene, it was this megafauna that exerted the greatest influence on the environment. If a 100 kg wild boar was capable of resculpting a forest, imagine the impact of a 6,000 kg mammoth. These mammals acted as heavy machinery: a convoy of bulldozers, ploughs and diggers that shaped the ecosystems where they lived. The prowess of such creatures can still be witnessed today, albeit on a smaller scale, in Africa, where elephants maintain the savannah by pulling up shrubs, pushing over trees and stripping trunks of their bark.[15] It was familiarity with these large modern-day herbivores that led the South African ecologist Norman Owen-Smith to first suggest, during the 1980s, that the combined impact of the Pleistocene megafauna 'could have been vastly greater than anything yet documented in Africa' – and that the loss of them would have unleashed a cascade of ecological consequences that still reverberate today.[16]

For, with a few exceptions, the megafauna did not persist beyond the dusk of the Pleistocene. Starting around 50,000 years ago, these creatures began to go extinct. Although debate has raged about the relative influence of humans and climate change in their demise, scientists have increasingly accepted that overhunting had something to do with it. The disappearance of the megafauna correlates closely with the arrival of migrating humans. Moreover, vulnerability to extinction appears to have depended upon how suddenly the hunters appeared. In Africa, where humans and megaherbivores had co-evolved, extinctions were less severe. The death toll was worse, but not immediately disastrous, in places where the megafauna had already met other *Homo* species, like Neanderthals, and had learned to exercise caution when confronted with a spear. In places like Australia, however, where *Homo sapiens* were the first hominin that these trusting mammals had ever encountered, their destruction was swift and merciless.[17]

There is absolutely no doubt that humans did hunt megafauna. Their weaponry grew increasingly sophisticated over time, progressing from wood to bone to stone. The points that they attached to their spears were the work of master craftsmen – perfectly balanced and as beautiful, in their way, as the figurines of lion gods and swimming reindeer that were being produced at the same time. They were absolutely lethal to boot. 'Solutrean points resemble the canines of sabre-toothed cats,' writes the Australian palaeontologist Tim Flannery, referencing one particular iteration of blade technology. 'Indeed, they may have killed in a similar way – by exsanguination.'[18] Dotted across the world are the remains of mammals brought down by such weapons: a mastodon with an antler projectile embedded in its rib, recovered from a pond in Washington State; a mammoth with an ivory shaft stuck in its shoulder bone, discovered in the Siberian Arctic; a giant ground sloth butchered in Argentina, the skull forcibly separated from the skeleton to allow the hunter access to its protein-rich brain.[19]

The extermination of these engineers had a dramatic impact on the environment. Imagine the ecosystem as a jigsaw puzzle. The top half is a clear blue sky and the bottom a busy beach scene: children build sandcastles, fathers read books and mothers lick ice-creams dipped in sprinkles. Lose a piece of the sky and the picture still makes sense – all the parts look roughly the same. Lose a piece from the beach, however, and vital information starts to vanish: the shape of the sandcastle, the name of the book, the flavour of the ice-cream. Your imagination can no longer fill in the blanks. The same is true of nature. When there is an abundance of animals doing the same role – eating the same fruits, dispersing the same seeds, uprooting the same soils – then certain species may be lost without too much disruption. When an animal has a unique job to do, however, then its absence threatens the integrity of the overall picture; the scene starts to make a little less sense. Nature becomes less resilient, more vulnerable to shocks, more prone to switching to another state entirely.

Take the mammoth steppe, a dry expanse of botanically rich grassland that, until the megafaunal extinctions, stretched across the earth's northern latitudes. The fertile soils sprouted a nutritious mat of grasses, herbs and willow shrubs that supported enormous herds of grazers. Towards the end of the Pleistocene, however, the vegetation flipped: the arid soils became waterlogged and a carpet of mosses threaded across the continents. Until the 1990s, it was argued that this switch was a result of a changing climate, including increased rainfall, which inundated the dusty soils of the steppe, and that this, in turn, led to the loss of the megafauna, which were unable to survive on the indigestible tundra vegetation that emerged. However, in 1995, a group of Russian scientists turned this hypothesis on its head: they proposed that it had been the loss of megaherbivores themselves that transformed the mammoth steppe from dry-and-grassy to wet-and-mossy. Either could have persisted within the existing climate, they argued, and it was only the presence of the mammals that had led to the dominance of the former state. By removing any incipient vegetation, the megafauna created the conditions for grassland to flourish. Absent the megafauna, however, the system broke down: shrubs gained a foothold, the water table rose, and mossy tundra began to take over.[20] A biome lost forever, with humans to blame.

The mammoth steppe is far from the only landscape to have flipped into a new state following the extinction of the megafauna. Across the globe, ecosystems collapsed as their architects were turned to meat. New ones emerged in their wake. The flower-filled rainforests of Australia were invaded by trees with spiky, leathery leaves, adapted to withstand the fires that began to blaze with more regularity without the animals to keep the foliage – the fuel – in check.[21] In the American Midwest, strange mixed forests of broadleaf and conifers emerged as browsing pressure decreased. In Europe, dense wildwood swamped the open, park-like landscapes without beasts to trim the shrubs.[22] The world may still have looked wild: there were no cities, no roads, no farms. But it is hard to see the post-megafaunal world as anything other than a human creation.

The loss of the megafauna did not only impact vegetation. Their extinction was the first in a chain of dominoes that wound through the rest of the animal kingdom. Smaller mammals and birds depended on the megaherbivores for maintaining the glades and grasslands they needed to survive. As trees and shrubs took over, such species would have found themselves restricted to the small pockets of suitable habitat that remained. Unable to migrate into more promising territory, they would have been vulnerable to human hunters on the lookout for a replacement for their megafaunal diet.[23]

It wasn't just humans who were forced to adapt to the sudden scarcity of meat. The flesh of the megaherbivores also fed the Pleistocene's mega-carnivores, which went extinct around the same time. One theory is that humans outcompeted these giant predators, who went hungry and ultimately perished as a result.[24] Mega-vampires also suffered from the paucity of prey. The blood of the megafauna sustained two species of giant bat, *Desmondus draculae* and *Desmondus stocki*. Unlike other vampire bats, they were unable to survive on the blood of the remaining mammals: both went extinct around the end of the Pleistocene. Scavengers, too, would have missed the corpses of herbivores that had died from natural causes. The loss of this ready supply of rotting flesh may have contributed to the extinction of the giant hyena and some of the larger species of vulture. The scavengers that did survive, such as the California condor, were forced to retreat to the coast, where they could still find a meal in the stranded carcasses of the ocean's giants.[25]

These days, there is great concern about the insect apocalypse – a crisis that has come to be defined by the falling number of bug splat-marks on car windows. However, the decline in invertebrates has been thousands of years in the making. Dung beetles have been around since the mid-Cretaceous period, when they fed upon dinosaur faeces. Later, they adapted to mammal faeces instead. But, with the extinction of the megafauna and the loss of their giant pats, the largest of these beetles finally went extinct – they could cope with an asteroid, it seemed, but not smaller piles of poo. Less mourned is the loss of the Pleistocene parasites. More than 400 species of the worm-like helminth died out alongside their mammalian hosts, as did one particular stomach botfly, the only testament to its existence a clump of twelve plump larvae discovered by a Russian scientist inside the frozen flesh of a mammoth in 1972.[26]

Astonishingly, plants are still reeling from the megafaunal extinctions. In the 1980s, tropical ecologist Dan Janzen and palaeoecologist Paul Martin realised that the lowland forests of Costa Rica contained numerous trees that appeared ill-adapted to their surroundings. Their fruits were large, fleshy and spiny and, with no way of dispersing beyond the mother tree, instead piled up in rotting heaps beneath the canopy. But the mystery, they wrote, 'disappears when interpreted in light of the extinct Pleistocene megafauna'.[27] These plants, they concluded, were ecological anachronisms, whose seeds were designed to be consumed and scattered across the landscape by giant animals with correspondingly giant digestive tracts. Many such plants went extinct in tandem with the creatures that dispersed them. Those that survived did so by forming relationships with different animal species, including livestock. Others endured through human domestication: bitter-tasting wild pumpkins, once dispersed by mastodons, live on today in sweeter form, cultivated to be baked into Thanksgiving pies and carved into jack o'lanterns at Halloween.[28]

The natural world has changed many times during the earth's 4.5 billion year history. The *T. Rex* did not walk the same landscapes as the woolly mammoth. But the transformation that occurred during the Pleistocene was different: not a neutral realignment of nature but a fundamental weakening of the natural order. Ecosystems we consider 'natural' today are haunted by the ghosts of the megafauna. Never quite able to shake the memory of their presence, it is as though the earth is poised for their return. But extinction is forever. In the untrampled trees and unspread seeds is a subtle reminder of how much was lost when we robbed the world of its greatest beasts.

Even so, the Pleistocene hunters were not the first humans to transform the world around them. A few years ago, archaeologists made a startling discovery, pushing back the boundary of human wildness further than ever thought possible.

The bones that began it were discovered in 1985, in an opencast mine in Neumark-Nord, East Germany. As the bucket-wheel excavator – a machine the size of a rollercoaster designed for scooping away vast quantities of rock and earth – bit into the quarry, the operators began to notice tiny white bands appearing in the mud. The onsite geologist hurried to the house of Dietrich Mania, an archaeologist at the nearby state museum, with a backpack full of deer bones. What Mania discovered, when he arrived at the site, was the most spectacular collection of Pleistocene fossils he had ever encountered – all about to be crushed in the rush for coal.

Cut off from fuel reserves in the west, the communist state was desperate for energy: the mining company wasn't stopping for anything. But the drivers weren't unmoved by the discovery. Over the next decade, they worked alongside Mania to rescue as many of the bones as they could. Sometimes, Mania would wait at the margins as the deposits of mud were removed and the remains became apparent. Occasionally, he received a phone call to say that a new skeleton was about to resurface, and he would rush to the mine in his grey Trabant car to see what could be saved.

Rescuing the bones meant sitting a couple of metres in front of the bucket-wheel excavator and working as quickly as possible to remove them from its path. The rescue mission was the work of many years; the machines worked day and night, which meant the archaeologists did, too. Mostly Mania worked alone, using spatulas and gardening tools to exhume the skeletons. Some bones were destroyed and many others fractured by the miners. When something really extraordinary materialised, the diggers sometimes agreed to change course to allow a little extra time. 'I remember very well, in March 1995, we had about three-and-a-half hours to excavate a complete elephant,' recalls Enrico Brühl, one of Mania's students at the time, who helped out during the summer months. The call came through on Brühl's birthday at 11 p.m. at night; the archaeologists were at the quarry, ready for action, at 6 a.m. the next morning. 'In that special case, they realised the elephant was so very well preserved that they moved the bucket-wheel over us and continued on the other side.'

The archaeologists discovered 98 straight-tusked elephants in total. There were also 44 bison, 19 aurochs, 17 rhino, 10 bear, 5 lion, 3 hyena and myriad deer – and that represents just a fraction of the 200-plus species that were recovered during the excavations. Even the contents of some of the animals' stomachs had been preserved. These animals had breathed their last around 125,000 years ago, during a period known as the Eemian or, more intuitively, as the last interglacial.

The Eemian began around 130,000 years ago and lasted some 15,000 years. The people who lived during this period were the lucky ones: the glaciers of the previous Ice Age had finally receded, and those of the next had yet to advance. The continent was green and fertile and still hosted the full range of megafauna; tribes of humans were still too small and spread out to have dented their populations. The temperature during this period was much like it is today. As such, scientists often consider the Eemian a portrait of how the continent might look now had humans not interfered to such a degree: the true 'natural' baseline for Europe's vegetation.

During the glacial period that preceded the Eemian, Neumark-Nord was covered by an ice sheet that, when it thawed, left behind a pockmarked terrain of channels and lakes. It was evidently a place to which humans were attracted. For around 2,000 years, a community of hunter-gatherers lived along the lakesides, where they spent their days collecting firewood, foraging for hazelnuts and spearing herbivores at close range. These humans, however, were not *Homo sapiens*. They were Neanderthals.

Ever since the discovery of the first Neanderthal skull in the summer of 1856, in a valley outside Düsseldorf, *Homo sapiens* have shown scant respect for their ancient cousins. At first, the skull's peculiar shape meant it was thought to have belonged to an ordinary human who had been suffering from rickets. It was only in 1864 that William King, a British geologist, proposed that it could have belonged to an extinct species of human – although 'the thoughts and desires which once dwelt within it never soared beyond those of a brute', he wrote. A decade later, the biologist Ernst Haeckel suggested renaming the skeleton *Homo stupidus*. The name never stuck. Instead, 'Neanderthal' itself has become shorthand for anyone deemed too brutish or dimwitted.

These days, the reputation of the Neanderthals has improved somewhat. New discoveries have revealed that they created jewellery and art, and possessed minds that were capable of abstract and symbolic thought – a cognitive function that, until recently, was believed to have been limited to *Homo sapiens*. Even so, the notion that they might have sculpted the landscape, intentionally and to their benefit, remained off the table. Before the investigation of Neumark-Nord, the oldest evidence for humans shaping the environment came from Lake Malawi. Here, remains of pollen, charcoal and stone tools in the dusty red soil suggested *Homo sapiens* had used fire to convert closed-canopy forest into the bushy open woodlands roughly 85,000 years ago.[29] The idea that another species, thousands of years earlier, could have similarly reimagined their surroundings seemed implausible – but the evidence suggests that this was exactly what happened.

Mining at Neumark-Nord ended in 1996. The archaeologists, too, packed up and went home. A few years later, however, the reunified government decided to clean up the mess left by the coal company and turn the abandoned quarry into a lake; today, daytrippers sail boats and pitch tents by the water, largely unaware of the elephant graveyard beneath their feet. The coal company provided funding for more systematic excavations before the flooding took place: another nearby basin, similarly rich in fossils, had been discovered but not explored during the earlier work. Without the pressure of bucket-wheel excavators at their backs, archaeologists uncovered many of the finer details that had been overlooked during the hurried efforts of the 1990s, including pollen, molluscs and charcoal. It was this data that allowed the researchers to start piecing together the scenery that would have framed the Neanderthals as they camped along the lakeside, knapping flints and eating elephant brain by the flickering light of the hearth.

What they found was a scene that shifted over time. The landscape was initially dominated by forest: first of birch, then of pine. There may have been occasional gaps in the canopy – the result of a storm or lightning strike – but the trees would have moved in quickly again. Then, suddenly, the ecosystem opened. Oaks replaced the conifers, lapping up the sunlight with their acorn-laden boughs. Herbs and grasses sprung up on the warm soil, soon followed by scrubby hazel. The shift in vegetation was accompanied by a sudden spike in charcoal. The forest, it seemed, had not reached the end of its natural life, but had burned to the ground. The proliferation of stone tools from around the same time suggests Neanderthals were the arsonists.

The landscape remained open for more than 2,000 years. It was only after people abandoned the lakeside that the forest returned. For Wil Roebroeks, the archaeologist who led the research, the implications were clear: having burned down the wood and set up camp in its ashes, Neanderthals spent the next two millennia beating back any encroaching trees. They were invested in the world they created. Their reasons for maintaining this open ground would have been similar to those of the Mesolithic hunters who prowled Europe more than 100,000 years later: the fresh grass would have attracted the species upon which they preyed and acted as a garden where nuts and berries could flourish.[30]

The settlement at Neumark-Nord is the oldest example of humans having a direct and lasting influence on the natural world. The discovery was unparalleled: there is no other reliable evidence of Neanderthals acting as sculptors of their surroundings. But absence of evidence is not evidence of absence.

Well-preserved records of early human activity are vanishingly rare. The discovery at Neumark-Nord was down to a combination of global politics and good fortune: had the glaciers of the next Ice Age extended a little further southwards, the records may well have been destroyed forever. The Neanderthals of Neumark-Nord were not necessarily ahead of their time, but rather camped in exactly the right place.

'What we are looking at, with this archaeological record in Neumark, is the refuse – the rubbish, the garbage – of the hunting and animal processing and plant-collecting activities,' Roebroeks tells me. 'The preservation conditions are atypical, but they document, we think, typical run-of-the-mill Neanderthal behaviour that just isn't fossilised a lot.'

The greatest question hanging over these findings is whether the Neanderthals burned the forest to the ground themselves, or simply made use of a landscape that had been opened through natural causes. The presence of charcoal alone cannot prove whether the fire was caused by humans or lightning, and the records are not precise enough to say whether the people or the fire came first. An earlier suggestion, put forward by Roebroeks' colleagues, was that herbivores, so abundant at the site, had broken up the forest with their trampling hooves, creating a glade that was subsequently adopted by the opportunistic nomads, who then set about lighting small campfires for cooking and protection.[31]

Such hypotheses are difficult to test, and the conclusions are by no means certain. When I asked Enrico Brühl – who went on to lead the second round of excavations after helping Mania during the original dig – for his thoughts on Roebroeks' findings, he appeared less than convinced. However, what makes Neumark-Nord particularly valuable, in Roebroeks' view, is the presence of two other nearby lakes, which were also exposed and excavated during the coal mining boom. All three basins sit in the rain shadow of the Harz Mountains, meaning that they have always experienced identical patterns of rainfall and temperature. They are close enough to one another that they would have been frequented by the same animals. Based on natural forces alone, the same ecosystems should have developed around each one – and, for a while, they did. Thick forests developed around all three lakes around the same time. Only at Neumark-Nord – the only lake with evidence for long-term human habitation – did that forest disappear.

Even so, Roebroeks stresses that there will always be some element of mystery when peering so far into the past. Yet, given that the chance of finding their campsites is so tiny, he believes that humans could well have been shaping the landscape long before the evidence appears for the first time. Once our ancestors had mastered fire, there was little to stop them using it to shape the landscape to their will. 'Given what we know about other animals, if humans could produce fire – and they could probably already do that 400,000 years ago – we are bound to find earlier evidence. I wouldn't be surprised,' says Roebroeks. 'For us, it's a big mental leap to accept they were doing it. But if you can produce fire, it's a small step.'

Either way, to find a truly natural baseline – a time when the natural world bore no marks of human influence – it is necessary to travel back to before the genesis of *Homo sapiens*. Which raises the question: has our species ever been wild at all?

TWO

Small and Beautiful Clues

The wildwood has always played an outsized role in the imagination. It is a place that hangs in that shadowy space between fact and fiction; at once glittering green labyrinth, hideout of hermits and hags, source of sustenance for hunter and peasant. As an ecosystem, it seems somehow unbound by time: the backdrop to everything from the earliest Britons to the arrival of the Roman Army to the chivalric quests of medieval knights. At least, that is the impression left by the many obscure works of fiction, history and travel-writing that I perused in an attempt to understand how people have historically perceived Britain's ancient landscapes.

To delve into the nineteenth-century literature on this topic is to realise just how deep the perception runs. Thus, the purpose of Stonehenge may have been to civilise 'the untamed children of primeval forests by sacred rites', mused one writer in 1847.[1] The thatched huts of ancient tribes must have been encircled by 'the grand and gloomy old forest, with its shadowy thickets, and dark dingles, and woody vallies [*sic*] untrodden by the foot of man', according to a historian in 1867.[2] Fast-forward to the Norman conquest and we hear of bands of Anglo-Saxons forced to 'lead a roaming life in the forests, living principally upon what they could procure by the chase', as another historian wrote in 1865.[3] The impression was reinforced among the general public by better-known writers such as Charles Dickens, who, in *A Child's History of England*, taught generations of students that England, upon the arrival of the Romans, 'was covered with forests, and swamps'. In *The Sword in the Stone*, published in 1938, the novelist T. H. White describes the twelfth-century forest that frames King Arthur's adventures as 'an enormous barrier of eternal trees, the dead ones fallen against the live and held to them by ivy'.

Such descriptions may have been somewhat romanticised for their audiences, yet the idea that England retained its primeval forest until the Middle Ages was not out of step with the scholarly literature of the time. As late as 1954, the eminent landscape historian W. G. Hoskins was arguing that, before the fifteenth century, 'England must have seemed one great forest ... an almost unbroken sea of tree-tops with a thin blue spiral of smoke rising here and there at long intervals.'[4]

Archaeologists, meanwhile, were less certain of the nature of the historical landscape. Around the start of the twentieth century, the profession started to consider the question seriously and systematically. The mystery was complicated by the long presence of humans on the island: there was no doubt in their minds that primitive tribes had inhabited Britain for thousands of years before the Roman invasion. Julius Caesar, who landed on the shores of Kent in 55 BCE, had described meeting its woad-painted occupants, clad in skins and surviving on a diet of flesh and milk. Beyond the written sources, the physical evidence for Britain's ancient people remained dotted across the landscape: a cryptic network of standing stones and strange earthworks, each requiring open land for their construction, attesting to a population that surely gathered and worshipped beyond the forest. Moreover, it was becoming increasingly apparent that agriculture had been a feature of the island for thousands of years – again, requiring a certain amount of open, unforested land.

In this context, the concept of an all-encompassing, uninhabitable wildwood was something of a conundrum. Britain's prehistoric inhabitants were understood to have been helpless bystanders within this fierce and verdant underworld. Cutting down trees was considered beyond their ken: they simply did not have the tools for the job. This forest was unlike the docile woodlands of the modern day, but rather a sprawling and unmapped mass of obstacles: the ground cluttered with rotting trunks, soils sodden by snowmelt and churned up by wallowing animals, the understorey tangled with holly and ivy. Then there were the predators: it was still possible, at this time, to encounter wolves and bears amid the trees. Our forebears, unskilled and unarmed, seemed to stand no chance at all.

'How could he fight the damp forest till he had metal with which to cut down its trees?' wrote Herbert J. Fleure, one of the first scholars to address the mystery, in a paper published in 1918.[5] He could draw only one conclusion: prehistoric Britain must not have been as uniformly wooded as had long been believed; certain areas must have been naturally free from forest.[6]

Upon this basis, Fleure produced a map showing what he calculated to have been the habitable areas of Southern Britain during the Neolithic period. Alongside the forests, he dismissed areas of swamp – 'probably haunts of ague' – as locations suited to settlement, as well as anywhere too high or craggy. Such restrictions automatically ruled out the majority of the island. However, there *were* a few areas remaining where he believed the pastoralists may have found land suited to the establishment of their settlements. These were the places where forest growth was

limited by either altitude or soils. They included coastal fringes, chalk downlands and upland moors – places connected by twisting trackways that, he surmised, may have had their origins in the tracks of wild beasts.

It wasn't until much later, with the arrival of metal tools, that Fleure believed people began to attack the forest. Even then, he suggested that any clearings would have been localised, and only for the purpose of providing winter shelter. Certainly, it seemed to him that they couldn't have made much progress into the wilderness by the time that Britain began to welcome its first visitors from the classical world. Fleure points to the Ancient Greek geographer Strabo who, writing around the turn of the first millennium, reported in his *Geography* that 'the island was mostly jungle'. It was only during the Roman period that he concedes humans may have acquired the skills to begin their descent into the forested lowlands.[7]

The idea that Britain's primeval residents were restricted to naturally open areas within an otherwise hostile forest was cemented by Sir Cyril Fox, one of England's foremost archaeologists of the twentieth century, in his 1932 work, *The Personality of Britain*.

Based upon a series of maps showing the locations of prehistoric monuments and remains compiled by his collaborator, Lily Chitty, Fox argued that Britain could be divided into two categories. On the light and permeable soils – chalk, limestone, sand and gravel – he found plentiful evidence for prehistoric occupation: these areas, he concluded, must have been either naturally bare or only lightly forested. On the sticky clay soils of the lowlands, on the other hand, there was not a monument to be found. Here, he wrote, would have been thick oakwood. To stray into this netherworld was to risk death and disease: sheep that found themselves in this damp environment sickened and died of liver-rot; humans would have been struck down by malaria caused by the marsh mosquitoes that bred in the stagnant pools. Bears, wolves and boars made their home among the trees, and the tangle of briars and brambles made escaping the charge of an aurochs impossible.[8] 'This type of forest,' wrote Fox, 'was definitely shunned by Man.'[9]

Fox conceived of the extent of Britain's ancient forests in squirrel miles. The story of this apocryphal rodent, crossing vast distances through the treetops without ever touching the ground, has been around for centuries, although I have never been able to uncover the true origins of the phrase.[10] The squirrel has scurried across the skylines of everywhere from the Iberian Peninsula to the Eastern United States. At some point, it made the leap into the axioms of British history, such that, in 1918, the architect Maurice B. Adams could claim that, during the Middle Ages, 'open country was unknown, and England was so densely wooded that it was said a squirrel could traverse the kingdom without touching the ground'.[11] In the conclusion to *The Personality of Britain*, Fox took a squirrel's-eye view of the landscape at the dawn of the Iron Age, around 2,500 years ago. The forest at this time 'was in a sense unbroken, for without emerging from its canopy a squirrel could traverse the country from end to end,' he wrote, echoing the old cliché. But Fox's squirrel would have needed to do some route planning to achieve this feat. His treetop journey would have been punctuated by patches of open land, where he may have observed humans milking cows and tending crops. It was in these forest-free stretches, according to Fox, that the civilisation of Britain truly began.

The Personality of Britain was hugely influential – not least because of Fox's failure to cite his influences, lending the impression that his work was more original than it really was (he would not reference Lily Chitty until the third edition). Even so, it quickly became clear that something was amiss. Monuments had a way of turning up on the wet soils where, according to his theory, they should have been absent. In 1936, the Welsh archaeologist William Francis Grimes pointed to an ancient burial chamber in Carmarthenshire, hidden by thick woodland at the bottom of a narrow valley, so overrun by vegetation that not even the local farmer knew its exact location. If trees could grow there in the present, Grimes supposed, there was no reason they should not have also grown there in the past – and, therefore, that the forest was not the barrier to habitation that it was widely supposed. 'At the moment the evidence suggests that the megalithic folk had established some sort of control over their surroundings,' he wrote, 'and I wonder whether their absence from the interior of the country is due rather to their own lack of interest in a type of country which may not have suited their mode of life, than an inability to overcome some at least of the natural obstacles which may have stood in their way'.[12]

Fox himself was largely unfazed by such inconsistencies. Rather than question his premise, he conjured up a prehistoric class-based system that explained away any outlying monuments, where the wildwood was the equivalent of the modern-day slum and open areas the preserves of the wealthy. 'All human communities, of course, throw off groups and families below the poverty line or their particular culture, who scratch a miserable living how they can in the less desirable areas,' he wrote. 'Evidences of such will certainly be found from time to time on the clays of the Lowlands; but they are negligible.'[13] Inconvenient findings, on this basis, could be dismissed as evidence of the dispossessed: of people who were forced to live *within* the wood, rather than people who were capable of clearing it.

A similar debate over the nature of the prehistoric landscape was playing out in Europe, centring on Germany. In 1867, the geologist Albrecht Penck made the idle suggestion that certain parts of the country, when first settled by Neolithic people, would have been as free from forest 'as the prairies of North America'. This view was formalised by the geographer and pastor Robert Gradmann, who proposed that the spread of farming must have been contingent upon the existence of an open steppe-heath ecosystem, where the soils could have been easily worked by their primitive occupants.[14]

The theories may have differed in their details, but the central premise remained the same: prehistoric people were governed by their environments. The stones and earthworks they left behind were signs not of their mastery over the forest but their subjugation before it. But, somehow, the physical evidence never quite conformed to the facts: our forebears stubbornly refused their consignment to the margins, demanding ever greater intellectual contortions from the scholars who studied them.

In Denmark, meanwhile, another man was on the case: not an archaeologist, but a botanist obsessed with small and beautiful clues. Unravelling the mysteries of the ancient landscape, it turned out, would turn not upon megaliths but microscopes.

When Johannes Iversen died in 1971, his colleague penned an obituary comparing him to Miss Marple.[15] Like Agatha Christie's famous detective, he wrote, Iversen had an uncanny ability to solve mysteries that had left everyone else stumped, thanks to an intimate knowledge of the players involved. In Iversen's case, those players were plants.

Iversen was born in Sønderborg, Denmark, in 1904. His parents were peasants, and deeply religious; his father, as a young man, had been a missionary for the Moravian Brethren, an obscure denomination of the Protestant Church. Iversen took this spiritual devotion and directed it towards the living world: those that knew him spoke of experiencing nature with a strange intensity whenever they were in his presence. He was gentle and serious, loath to pick the flowers he studied, preferring to unwind them tenderly and observe them *in situ*.

After completing his schooling in Germany, Iversen studied botany at the University of Copenhagen. In his academic work he was always meticulous – sometimes frustratingly so – writing only in pencil the better to erase, and submitting upwards of twenty drafts of his manuscripts. 'The pleasure of reading Iversen's papers,' commented one of his colleagues, 'is like that of listening to eighteenth-century music.'

It was this forensic attention to detail that made him perfectly suited to solving the riddles of the ancient landscape that had so bamboozled Britain's most eminent archaeologists. However, it was towards the great unsolved questions of Viking history that he would first turn his attention.

Iversen had just taken on his first job, as a part-time assistant at the Geological Survey of Denmark, when an invitation came through. For some years, the National Museum of Denmark had been excavating the Viking settlements of Greenland, and they wanted the young botanist to join them on their next excursion. The Vikings had settled the island around 1000 CE, establishing farms and settlements along its fjords. Four centuries later, they vanished. Why they left, or what forced them out, no one could tell. One theory was that the onset of the Little Ice Age – a period of cooler weather starting around the fourteenth century – had caused its communities to perish en masse.

But Iversen had another idea. Perusing the ruins of Vesterbygden, one of the old Viking settlements, he noticed that the Norsemen had left traces of their presence in the soil: a thin layer of charcoal, where the original vegetation had been burned to create fresh pasture for livestock. Just above the charcoal was a dense blanket of fossilised pupae and scaly hairs left behind by the great brocade moth. While climate change may have caused the slow decline of the Vikings in their new territory, Iversen believed it was this sudden infestation that pushed them over the edge, the moths descending like a musty curtain over the springtime fodder and causing their cattle to starve to death, leaving the settlers vulnerable to attacks from the locals. 'Like a plague of Egyptian locusts, a larval attack means an acute catastrophe,' he wrote. 'A single year would have been enough to completely ruin Vesterbygden and break the resistance to the hostile Eskimos, who, as hunters and fishermen, remained untouched by the disaster.'[16]

Next, Iversen turned his mind to the Vinland Question. The controversy – about as fierce as they come in the esoteric world of Old Norse literature – centred on whether the Vikings had really landed in North America some 500 years before the arrival of Christopher Columbus, as claimed within two thirteenth-century texts: *The Saga of the Greenlanders* and *The Saga of Erik the Red*. Both sagas describe voyages by Greenland's early Viking settlers to a place called Vinland. Neither text, however, is a reliable historical document: both are based upon ten generations of storytelling rather than first-hand accounts from the seafarers themselves. The debate over the veracity of the sagas had raged on for around a century, and, with no physical evidence one way or the other, there was no end in sight. At least, not until Iversen showed up.

One day, while wandering the abandoned farmsteads of Greenland's Godthaab Fjord, the botanist came across a small blue flower, *Sisyrinchium montanum*. It was a long way from home. The species had never before been found in the Arctic. Rather, it grows over 1,000 km away across the Labrador Sea – in Canada.[17]

Vinland is most commonly translated to 'land of wine', but it can also mean 'pasture land'. *Sisyrinchium montanum* is a flower that grows in meadows. Its seeds are easily transported within the coats of cattle or bundles of hay, but are otherwise ill-adapted to long-distance travel: it was unlikely they had made it to Greenland under their own steam. In the presence of this flower, so far from its homeland, Iversen found the most convincing evidence yet that the Vikings had made the journey out west. The seafarers – seasoned farmers – must have settled in Canada, fattening their cows in the meadows where the species grew naturally, before returning home with their animals in tow, the seeds hitching a ride in their fur.[18] In 1960 the discovery of timber buildings filled with Viking relics at the tip of Newfoundland would confirm the answer to the Vinland Question once and for all: the Vikings had made it to North America. Iversen was right.

Old larvae and tiny flowers may have helped the young botanist to rewrite the history of the Vikings, but to investigate the history of the prehistoric landscape more widely Iversen would have to turn to even smaller clues: fossilised grains of pollen, thousands of years old.

Plants create pollen to reproduce. The grains spread on the wind, or on the body of an insect, seeking a female to fertilise: those that fail spread across the ground in a fine layer of dust. Their tough outer walls are composed of sporopollenin, a biopolymer that is almost entirely resistant to decay, which means that, in the right conditions, pollen can survive intact for millions of years. The grains of each species are unique, variously resembling conkers, coffee beans, sea urchins and brains. Over time, these build up to create a rough archive of the landscape as it would have looked, season by season, in times gone by; a microcosm of the weeds and woodlands that have long since vanished.

The study of pollen, known as palynology, was relatively new in Iversen's time. The Swedish geologist Lennart von Post had outlined its potential for the first time in a lecture in 1916. Applying the knowledge was not always so easy: when Iversen began his investigations, microscope lenses were still so expensive that the Geological Survey of Denmark kept them in a safe.

Ultimately, it was the pollen buried within the salty mud and dried-up lakes of Denmark – not the mysterious megaliths of the English downlands – that would reveal the true nature of Europe's ancient landscape. The story that Iversen extracted from those grains went more or less as follows.

After the glaciers of the last Ice Age had finally retreated from Europe, a fierce wildwood did spread across the continent: a continuous closed-canopy forest, composed of oak, elm, ash and lime, in which the darkness gave way only to swamps. This tangled world was inhabited by hunter-gatherers, who 'had their paths as the animals had their tracks', wrote Iversen, echoing the views of the British archaeologists of the time.[19] But then he diverged from the accepted narrative. After just a few millennia of dominance, he wrote, the wildwood retreated, and a shrubby mat of birch and hazel sprung up in its place. The shift was sudden; abrupt. The pollen laid down at this time was accompanied by a thin layer of charcoal: a sure sign of fire. The indomitable forest had burned to the ground.

The fire-starters, wrote Iversen, were humans. Farmers. Where the wildwood vanished, there was a sudden increase in weeds, including mugwort and plantain – species that typically follow in the footsteps of agriculturalists as they spread around the world – as well as signs that cereals were being grown intentionally in the ashy soils that followed the burn. The fresh green shoots of the emerging trees, meanwhile, would have provided excellent fodder for their cattle. The clearance was no accident,

Iversen surmised, but the work of skilled labourers, armed with stone axes and flames. Of people both willing and able to bend nature to their will.[20]

Iversen published his findings in 1941. The explanation was so elegant, the ecology of it so persuasive, that it was immediately accepted by botanists, although the march of Hitler's armies through Europe disrupted its circulation. 'I recall vividly the thrill with which I received clandestinely through Sweden, in the dark days of the war, his monumental paper,' recalled the famous English botanist, Harry Godwin, some years later.[21]

Archaeologists, on the other hand, remained unconvinced. The idea that primitive man could so completely transform the landscape was such a leap from the prevailing theory that they were reluctant to accept it. So Iversen and his colleagues embarked upon the only logical course of action: they would clear a forest themselves, using the very same tools that those farmers had held in their hands some 6,000 years earlier.

On a September day in 1952, an unlikely team of archaeologists and lumberjacks set up camp in Draved Wood in Southern Denmark. With its mix of ancient oak, lime and alder, it was the closest approximation that Iversen could find to the original wildwood. Iversen's colleagues, Jørgen Troels-Smith and Svend Jørgensen, were to adopt the role of Neolithic axe-men, alongside two professional woodsmen. It was Troels-Smith who insisted that they use authentic Stone Age blades, persuading the National Museum in Copenhagen to lend them the flints from their collections. The blades were then hafted to replicas of a wooden shaft that had been recovered from a Danish bog, crafted by the museum's joiner, to complete the axe.

The archaeologists gained permission to cut down 2 hectares of the forest. The team had perfected their method during the depths of the Danish winter. The most effective way to fell a tree with a Stone Age axe seemed to involve striking the trunk with the power of the forearms alone, using small quick blows to cut a triangle out of the wood, weakening the structure until the entire tree could be toppled to the ground.

Come spring, they were ready to take on the forest in earnest. Together, Jørgensen, bare-chested beneath his overalls, and Troels-Smith, rarely pictured without his waistcoat and pipe, swung at the trees until their bodies ached. After supper, the three archaeologists would stroll throughout the darkening forest, so deep in discussion that they were sometimes still awake when dawn broke and the morning dew began to drench the grass. 'I have twice in my life reached the furthest limits of my endurance,' remembered Troels-Smith in a lecture delivered some years later: 'physically while felling trees in Draved Forest with a ground flint axe and mentally while together with Iversen preparing "Pollen morphological definitions and types."'

The archaeologists, it turned out, were better at clearing the forest than the professional woodsmen, who struggled to unlearn a lifetime of habits tailored to modern tools. The Stone Age axes were efficient when used correctly, but otherwise delicate, and the lumberjacks damaged several of the ancient blades with their powerful blows. Troels-Smith, on the other hand, cleared his section of the forest using a single flint, unpolished since it was chipped into shape by its primeval craftsman, back when the wildwood was still standing.

In fact, clearing the ancient woodland proved remarkably easy. Draved fell quickly before the axe. In one day, three people managed to chop down 500 square metres of forest. Considering that the Neolithic woodsmen were likely far stronger and more experienced workers than the archaeologists, and equipped with sharper tools, the notion that the wildwood posed an impassable barrier to human occupation suddenly seemed impossible to sustain. The first farmers were not passive subjects of the natural world: they were its architects.

After the cut came the burn. The archaeologists used a method of farming with fire, known as swiddening, that was still practised in rural Finland in the twentieth century, and which was thought to have been an analogue for how clearance would have taken place across primeval Europe. They spread brushwood and branches from the felled trees across the remaining scrub and used flaming torches of birch bark to set them alight. It was a resounding success. Within a few days, the burn was complete: a field of fertile ash lay where there had once been forest, ready to be sown with crops.

The next stage of the experiment showed why the first farmers were so keen on fire. The prehistoric seed varieties failed to take on the unburned ground, but the sweet, smoky soils of the razed patches produced a luxuriant crop – for one year, at least. After that, the harvest failed and a rough carpet of flowers, herbs and mosses began to take over. At this point, a real Neolithic community would have abandoned their plot and cleared a new section of the forest – it being easier to start from scratch than to tackle the regrowth – thus penetrating ever deeper into the wilderness, creating bit-by-bit the landscape of Europe as we know it today.

The clearance of the wildwood took place unevenly across time and space. But cleared it was. More recent research has shown that, by the end of the Bronze Age, Europe had already lost a tenth of its temperate forests, with much of that clearance concentrated in Britain. Deforestation continued apace through the Iron Age.[22] The notion that the Romans encountered unending wildwood when they alighted on the island's shores is pure fantasy; romantic portrayals of medieval castles nestled within an ocean of treetops are even more far-fetched. When Europeans mourn the loss of the world's tropical rainforests, they do so from within the graveyards of their own devastated woodlands.

Which leaves the question of why Cyril Fox failed to find any evidence of prehistoric habitation in certain parts of Britain. The answer lay in the soil – just not in the way he supposed. The absence of monuments was not down to the presence of wildwood but the plough. This machine, which prepares the fields for crops, has the side effect of returning old earthworks to the ground, destroying the evidence of the past. The thin soils of the chalk downlands and the boggy peat of the uplands have always been difficult to cultivate, meaning that monuments here were more likely to survive the intensification of agriculture that has occurred over the last few centuries. The clustering of remains in these places was not because they were more attractive to prehistoric farmers, but because the farmers of modern times had shunned them: a cryptic map not of Britain's ancient settlements but its most fertile soils.

These days, the ability of prehistoric farmers to clear vast tracts of forest is no longer in doubt. But the mysteries of the ancient landscape have still not been completely unravelled. Today, the debate over the wildwood is fiercer than ever. It concerns not whether Europe's early farmers were capable of deforestation, but whether there was ever a primeval forest to fell in the first place.

Pollen analysis is excellent at showing whether a species would have been present at a certain point in time. However, it cannot describe how the vegetation would have been structured with any level of precision. Pollen is light and easily dispersed by the wind; a grain may settle miles from the tree that produced it. From this evidence alone, it is hard to determine the arrangement of the lost vegetation: were the trees squeezed together in tight squadrons, or did they spread themselves more patchily across the landscape, extending their gnarled boughs in the space around them? The wildwood was supposed to be something like the former: a cramped and dusky world where the summer sky would have been blocked by leaves.

In 2000, however, a Dutch biologist called Frans Vera put forward a ground-breaking hypothesis: the closed-canopy forest was a myth. Europe's temperate lowlands had never been possessed by dense and tangled wildwood. Instead, the landscape that emerged following the last Ice Age would have looked more like a modern-day parkland or wood-pasture: a shifting mosaic of groves, grassland and scrub, shaped by wandering herds of herbivores – the aurochs, elk, horses, bison, boar and deer that would have been so abundant at that time. The scientific orthodoxy until that point had been that such animals could only ever have been present in small numbers: it was believed that the forest could not have survived the pressures of so many mouths and hooves. Vera's point, however, was that ecologists had led themselves astray by taking the existence of closed-canopy forest as their starting point. He hypothesised that large numbers of herbivores had been present in the landscape and that they did prevent the forest taking hold.

The only seedlings that could have survived in the presence of so many herbivores, according to Vera, were those that sprouted within the protective cages of the shrubbery. From beneath the spiky stems of blackthorn, roses and wild fruit would eventually emerge small groves of trees – sylvan islands among grasslands kept tightly trimmed by beasts. Ultimately, however, the oldest trees within those groves would collapse and die, and the resulting glades would fill with flowers as the sunlight poured in. The flowers would attract the herbivores, which prevented the forest from returning. Over time, however, the scrub would creep back and the cycle start over. Vera believed it was this evolving mosaic of habitats – not the razed remains of the wildwood – into which the first farmers scattered their grains of emmer wheat and einkorn.

Rather than relying on pollen analysis alone, Vera grounded his theory in how trees behave in the present day. Oak and hazel are abundant in the pollen records of the early Holocene. Both are trees which thrive in well-lit conditions and fail to regenerate beneath the shade of the canopy – their presence tends to point to sunny pastures rather than shadowy wood. The apparent absence of pollen from grassland species, meanwhile, did not mean that grassland was absent. Rather, Vera argued that the flowers that might have been expected in such a habitat would have fallen victim to the herbivores themselves, their tasty blooms shorn off before they had had a chance to spread their dust. Modern-day parkland landscapes, Vera observed, tend to produce pollen diagrams that are remarkably similar to those from the post-glacial period.

Beneath Vera's ideas lay a disquieting notion: that the natural world has drifted so far from its origins that we no longer know what counts as natural at all. His hypothesis challenged one of the fundamental principles of conservation: that restoring the pristine baseline of Europe means bringing back the forests. Conservationists were forced to confront the possibility that they had been focusing on the wrong ecosystem for all this time – and, in doing so, that they had been harming the species they sought to preserve. Vera argued that, not only was wood-pasture the original ecosystem of Europe, it was also the more diverse. The variations within this mosaic – its subtleties of light and shade, depth and edges, youth and rot – made it attractive to a wide range of species: wild fruit, lichen, songbirds, raptors, butterflies and beetles. These, he contended, were the rightful occupants of the landscape; and, by cloaking the continent in artificial darkness, conservationists were pushing them to the brink.[23]

It has been more than two decades since Frans Vera published his hypothesis. Though his ideas remain controversial, the principles are already being put into practice.

While some conservationists remain wedded to reviving the forest, others have wholeheartedly embraced semi-wild herbivores in their efforts to restore Europe's landscapes to a more natural condition. In the UK, the best-known attempt to recreate the ancient wood-pasture is at the Knepp Estate in West Sussex, where pigs and cattle – proxies for wild boar and aurochs – freely roam the land, mixing with herds of ponies and deer.

Isabella Tree, who documented the transformation of Knepp from intensive farmland to modern-day wilderness in her best-selling book *Wilding*, characterised Vera's hypothesis as 'common sense to anyone with practical knowledge of trees' and criticised the scientists who expressed doubts over its veracity. 'But the world of academia is a strange, sometimes counterproductive and often sluggish place,' she wrote. 'Where one might expect it to be open and responsive to new thinking, it can be oddly conservative and resistant to new ideas.'[24]

This implies that opposition to Vera was something of a knee-jerk reaction: the threatened establishment rejecting the radical newcomer. In fact, there was widespread interest in the hypothesis when it was first published, and palaeoecologists were keen to test the idea, filling the gaps that remained in the thesis using the tools at their disposal – not only pollen records but also fossilised beetles and other proxies for lost vegetation. By and large, this research has rebutted rather than reinforced Vera's findings. Most scientists demurred when I asked them for their take on the debate while researching this book. Common sense, it seems, is not necessarily the best guide to the primeval world.

One study, for instance, examined pollen records from Ireland. As we saw in the previous chapter, temperature fluctuations during the Pleistocene had stripped the island of most of its large herbivores. Without herds of bison, aurochs and horses to keep the forest at bay, light-loving species like oak and hazel ought to have been scarce. In fact, these trees turned out to be just as abundant in Ireland as across the rest of Europe, contradicting the notion that their regeneration was contingent upon herbivores keeping the forest open, suggesting that they may indeed have been part of the wildwood vegetation.[25] Further research has shown that prehistoric oaks may actually have been more shade-tolerant than those of today, with the introduction of powdery mildew, a fungal disease, to Europe in 1907 preventing modern oaks from regenerating beneath the shade of their own canopies.[26] Another suggestion is that gaps in the canopy, caused by fire rather than by mammals, provided all the light that these ancient oaks needed to thrive.[27]

Fossilised beetles have also shed some light on the controversy. Beetle species are often closely associated with particular habitats – even particular trees – meaning that their preserved remains can reveal what kind of vegetation was present when the bug was alive. An abundance of woodland specialists suggests that the landscape was forested; lots of grassland specialists imply plentiful open space; many dung beetles denote a lot of dung, and therefore a lot of herbivores to produce that dung.

Britain abounds with the remains of ancient beetles. One study tested Vera's theory by mapping the lives and deaths of these insects across the island. The results suggested that, following the retreat of the glaciers, the landscape was fairly open – at first. The lack of dung beetles suggests that this openness was down to the climate rather than herbivores. Over time, however, there is an increase in beetles associated with shade-tolerant trees, such as lime and elm, indicating the coming of the wildwood. Dung beetles were present only in small numbers at this time, and mostly located within natural openings caused by the migration of river channels, suggesting that herbivores were scarce on the whole. Reading between the lines of broken shell and shattered wing, it seemed that clearings would have been a persistent feature of the landscape, even when the wildwood was at its thickest, but that there was nothing like the scrubby wood-pasture that Vera had envisioned, nor anything to suggest that herbivores had a significant role in shaping the structure of the forest.[28]

Indeed, the extinction of the Pleistocene megafauna has been the thorn in the side of the Vera hypothesis. His theory focuses on the landscapes of the early Holocene. But, as we saw in the previous chapter, the largest of the ecosystem engineers had already vanished by this point. During the last interglacial, the straight-tusked elephant was Britain's largest herbivore. At the start of the Holocene, it was the aurochs. Over the same period, the median weight of the largest herbivores shrank by around two-thirds.[29] Their ability to shape the natural world shrank accordingly.

Before the megafauna went extinct, however, evidence suggests that Vera was most likely correct: that, for as long as Europe retained its full suite of hooves and tusks and teeth, the continent's ice-free landscapes were bright and open, dotted with trees rather than swamped by forest. A study published in 2023 painted an open, if varied, picture of Pleistocene vegetation. Based upon pollen records from almost 100 sites across Europe, scientists deduced that, during the last interglacial, around 130,000 years ago, the continent would have been mostly open or only lightly wooded, particularly in mild oceanic climates where herbivores could thrive. Closed-canopy forests dominated only in the mountainous Alpine regions, where steep-sided slopes and rocky terrain meant that herbivores were relatively uncommon.[30]

It seemed that sunlit wood-pasture may well have comprised the natural vegetation of Europe after all. But finding it meant delving deeper in time, back to when mammoths and rhinoceroses and other such animals still roamed the continent. The wildwood wasn't a myth, but nor was it natural. The dense forests that emerged at the dawn of the Holocene were an aberration: not the majestic climax of a pristine landscape, but the overgrown weeds of a world stripped of its beasts.

The Vera hypothesis has been controversial not only because of the difficulty of proving it one way or another, but because it has widespread implications for nature conservation. Livestock, particularly sheep, have typically been regarded as enemies of biodiversity, with overgrazing preventing the regeneration of forests. However, when livestock are reframed as replacements for lost megafauna, then they are no longer inimical to wilderness: they are essential to it. Vera's ideas have helped to rebrand farmers as conservationists. If their livestock are subsequently sent to the slaughterhouse and sold as premium fillets, then so be it. This has been the case at Knepp, which sells 75 tonnes of its 'wild range' meat every year.[31] That is not to deny the benefits to nature that have accrued on the estate, but whether the land amounts to a genuine approximation of wilderness or a glorified farm remains hotly debated.

Vera himself, in his original thesis, was explicitly hostile to agriculture. Farms may superficially resemble the ancient landscape, he wrote, but they have not enriched biodiversity. Rather, they compete with it by devoting huge swathes of the earth's surface to a fraction of its species. The open spaces that exist today are but a pale analogue of the teeming, heaving, steaming wood-pastures of old, where thorny scrub has been replaced by hedges and metal fences, and wild aurochs by domestic cows modified to yield the maximum amount of milk. 'The pressure of selection by agriculture on nature is present all over the world and leads to simplification and depletion of natural ecosystems everywhere,' he concluded.[32] While the restoration of Europe's primeval wood-pasture does require the reintroduction of herbivores, Vera argued that these animals should be rehabilitated into the wild rather than kept behind fences; brought back not because they are profitable but because they belong.

That is easier said than done. The ultimate testing ground for Vera's hypothesis has been a patch of reclaimed land in his own country, the Netherlands, called the Oostvaardersplassen. During the 1980s, with Vera's help, herds of grazers were brought into the reserve, where – as predicted – they set about sculpting the landscape. Over time, their numbers exploded: there were no predators to pick off the weak, and tall wire fences prevented them from migrating when their food ran out. Eventually, the herbivores exceeded the carrying capacity of the land. Thousands starved to death.

The sight of so many carcasses, less than an hour's drive from Amsterdam, divided Dutch society. Some, like Vera, defended the deaths as nature's way – no worse, certainly, than the industrial slaughter of farm animals. But others objected to the sight of so much avoidable suffering. There were funeral marches and minutes of silence for the dead animals; protesters brought crosses and coffins to the reserve. Vera and his family received death threats.[33] Eventually, the outrage led to a change in policy, with excess grazers being shot and sold to consumers, leading Vera to ultimately turn against his own project and label the Oostvaardersplassen a 'meat factory'.[34]

The story demonstrates the complexities of attempting to recreate the primeval landscape in the modern world. The unsettled debate over the original state of nature in Europe has never prevented conservationists from acting on what they believe – or want – to be the answer. Sooner or later, however, any project attempting to reinvent the wilderness will butt up against the boundaries, physical and psychological, of modern society. Not everyone wants to dial back the clock – not even within a small, fenced space that they never need visit.

But these qualms should not lead us to disregard the past entirely. We may never know the true nature of the ancient landscape, and there are too many missing pieces to attempt any kind of authentic reconstruction. We may only ever catch glimpses of this strange and bountiful time. When it comes to enhancing the present, however, that may be enough. What is clear is that the primeval canvas of Europe contained something of everything. It was a world forged and undone by the elements, the animals, and the cycles of death and regeneration. These are processes we can – imperfectly – restore within the landscapes of Europe today. The wilderness of prehistory may be out of reach, but its life force is still within our grasp.

THREE

Return of the Native

One day, around 6,000 years ago, a hunter abandoned a longbow on a mountain plateau.

He had been stalking deer at close range, hoping to spear a weakling as it grazed upon the shrubs and mosses of the open heath. Made from a single stave of yew, the bow should have been strong and supple. Its wood was straight and unknotted, probably sourced from a tree grown specifically for crafting quality weapons. The hunter had gone to some lengths to procure it: the absence of yew trees from Scotland at that time meant that the bow would have been imported from Cumbria or perhaps Ireland.

And yet, when the hunter pulled the string taut, the wood snapped. Useless. The deer seized its chance to flee. Cold and frustrated, he flung the bow to the ground and wondered what to do about dinner.

Venison was off the menu, but perhaps he would not have to go hungry that night. There was still a feast to be found in the wooded valley below. Hazelnuts and cherries hung off the trees, and blueberries studded the forest floor. The river flashed silver with salmon and trout, and wildfowl waddled unsuspectingly along its banks. Perhaps the hunter was even relieved to abandon the windswept precipice for the shelter of the trees. Or perhaps he was gored by a wild boar, savaged by a bear, or trampled by an aurochs. The glen had not yet been robbed of its beasts.

We will never know the fate of the hunter that day – but we do know what happened to his bow. For the next six millennia, it remained buried where it landed, preserved by the peat even as the landscape around it was utterly transformed.

Not long after the hunter departed, the first farmers arrived on the scene, bringing fire to the valley. The peat dried out in the heat, causing purple heather to explode across the plateau, replacing the bogbean and grasses that had thrived on the once soggy ground. But the farmers refrained from clearing the forest in its entirety. The impact of humans was gentle at first, and for thousands of years a tangle of hazel, alder, birch and oak continued to dominate the hillsides.

That all changed with the arrival of the Cistercian monks. In 1136, a group of men from the Yorkshire abbey of Rievaulx set up a new foundation in the Scottish borders, not too far from the valley in question. Melrose Abbey would turn out to be the first of many such establishments. For these settlers, the hills were not just somewhere to contemplate the holy spirit but a place to keep their vast flocks of sheep. The Cistercians were master wool-traders, their wealth built upon the backs of their livestock, and their arrival marked the beginning of the end of the wildwood. Over the following centuries, the rich forest was reduced to a grassy wasteland.

The Cistercians eventually left. The sheep, however, remained, now in the hands of traditional farmers, and supplemented by a population of feral goats. Constant grazing prevented the woodland from regenerating, ensuring the grass remained a trim monoculture of green. Later, Victorian collectors, in the grip of fern-fever, would pluck away at the specimens that, thanks to their remote locations, had survived the sheep. In 1857, a botanist called John Sadler described his attempt to gather *Woodsia ilvensis* from an overhanging ledge. 'The plant does not seem to be very plentiful where we visited,' he wrote, 'five small tufts being all we observed, of which we took four, leaving the other as an "egg in the nest".' It was the last time the fern was found in the valley.

Over time, the peat upon the plateau degraded to such an extent that the hunter's longbow began to peek out from its ancient resting place. In 1990, it attracted the attention of a passing hillwalker. The passage of time had elevated its status from broken weapon to archaeological treasure: the oldest longbow ever discovered in Britain. It was christened the Rotten Bottom bow after the bog where it was recovered. Conservationists quickly realised that it could hold the key to deciphering the environmental history – and promoting the restoration – of the hills where it had so long lain.

Environmentalists owe a surprising amount of their understanding of the past to alcohol. When Johannes Iversen was researching the history of the wildwood, several decades earlier, the Carlsberg Foundation funded his work. This time, it was cider to the rescue. Following the discovery of the bow, the National Trust for Scotland persuaded Strongbow to fund a study into the layers of pollen preserved in the peat where it was found, on the grounds that the artefact bore a striking resemblance to the company's own logo.

The peat, more than three metres deep in places, contained the complete story of the wildwood, from germination to bitter end. In the fossilised pollen was mapped the spread of the first trees after the end of the last Ice Age, as glacial debris transformed into fertile soil. Birch, the ultimate pioneer species, appeared first, its silvery bark a sylvan approximation of the glaciers it replaced. Next came hazel and juniper, and then pine. The passing of another 1,000 years saw the arrival of elm and then oak. Another 2,000 years and alder, strung with catkins, sprung up on the wetter soils.

Slowly, slowly, a place once frozen and hostile grew temperate and green. By the time the hunter arrived on the scene, the valley was richer than it had ever been before, a dewy maze of rowan, hawthorn, holly, cherry and aspen. This was the wildwood's most dazzling scene – a brief moment of brilliance, suspended between ice and sheep.

To most people, a lost world. To a pair of local zoologists, a challenge.

Peebles is an old market town in the Scottish Borders. Its coat of arms bears the motto: *Contra Nando Incrementum*. In English, 'There is growth by swimming against the stream.' It is a reference to the salmon that migrate up the River Tweed every year to breed, though it could equally apply to Philip and Myrtle Ashmole, whose quiet lives on the outskirts belie several decades of shared adventure.

The couple have spent years on the world's remotest islands in pursuit of a deeper knowledge of the natural world – usually, but not exclusively, focusing on seabirds. In the early sixties, you might have found them on Christmas Island, collecting regurgitated food from the red-tailed tropic-bird and the phoenix petrel. During the nineties, they discovered a new genus of spider while exploring the barren lava habitats and caves of Ascension Island. During one stint on Saint Helena, Philip discovered the pincers of a large earwig that was thought to be extinct, providing brief hope for the resurrection of the species. (It actually did go extinct a few years later.)

It was the decade they spent in America, however, that gave the couple a real taste for wilderness. Upon returning to Britain, they decided to settle not in Edinburgh, where Philip had a job at the university, but rather an hour's drive away, in a little house on the edge of the hills. A disappointing decision, as it would turn out: the environment around Peebles was as ecologically denuded as any city suburb. To them, the landscape resembled a skeleton stripped of its flesh: grass where there should have been forest, timber plantations where once there was wildwood.

But there were places that hinted at what had been lost. Trees clung on, like stranded ghosts, in the most rugged corners and steepest ravines – in the places, in other words, that were inaccessible to sheep. It was one such relic – an islet of birch trees, moated by the waters of Loch Skene – that first gave Philip his grand idea. He would bring back the wildwood. And he would do it right there: on his own doorstep.

Over the days that followed, he sketched out the details of his vision. His wildwood would be a place where it was possible to step back in time; to experience nature as it existed before humans hacked away at its roots. The native fauna – the lynx and the bears and so on – were obviously out of reach, but he could see no reason why the full suite of primeval vegetation shouldn't be restored to one of the many naked valleys of the Borders. He pictured a rich tableau of broadleaf woodland, montane scrub and upland heath, tailored to the altitude of the hills, sweeping along the riverbanks, across the slopes and up to the skyline.[1]

Back in Peebles, there was plenty of enthusiasm for the idea. The Ashmoles set up the Wildwood Group, which first met in October 1995. From there on, its members gathered regularly, discussing their plans over pub food and beer. The biggest obstacle to their vision was that none of them owned any land. If the wildwood was to be reborn, they would need somewhere to put it. Preferably somewhere very large. Philip was determined that the forest should not be some glorified arboretum but rather an immersive experience, where natural processes could unfold at their own pace, in their own direction, without instantly butting up against artificial boundaries.

In Scotland, where the majority of land is owned by large private estates, that was not going to be easy. Valleys rarely come up for sale. When they do, the costs run into the millions. But just as the project was starting to look like an impossible dream, one of the group happened to get talking to a neighbour. His cousin, he said, was a local sheep farmer who had just bought a big farm near Moffat, and was sure to sell them some land.

That farm happened to occupy the very same valley where the longbow had remained buried for so long – and it turned out that the farmer was indeed willing to sell.

The land, now going by the name of Carrifran, was the perfect place for the new wildwood. Surrounded by steep slopes, with the rainfall trickling down from all sides into a winding river, it had the feeling of being somehow set apart from the rest of the landscape. But what was truly unique was its pollen record. Not only did the Wildwood Group have their land, they also had a blueprint: an inventory of the species that they would need to grow to bring the ancient forest back from the dead.

Fidelity to the past is the guiding principle of the Carrifran Wildwood project – and that, perhaps, is the most audacious element of an already extraordinary plan. Such stubborn insistence on the romantic over the attainable is unusual in nature conservation today.

In fact, the official policy in the United Kingdom is to keep the landscape pretty much in stasis, frozen at the current point in time, regardless of the losses of the recent past. Sites of special scientific interest (SSSIs) are the jewels in the crown of the country's network of protected places. They are selected based on the presence of something that is deemed particularly valuable – a certain bird or flower, or a prime example of a habitat, for instance – which is then preserved at all costs.

That may sound sensible, but in reality it means that these sites, like Peter Pan, may never grow up. There is scant consideration of how the protected feature came to exist at the site – whether that habitat is, in fact, the degraded remnant of something else – and the natural dynamism of the ecosystem is resisted. This method of conservation was established by an Act of Parliament in 1949, which means that all SSSIs have come into existence in the last seventy years or so. Those decades are treated as a golden age for nature, rather than what they really are: an era of unprecedented destruction.

Carrifran sits within the Moffat Hills SSSI, which meant that, in the eyes of the Scottish government, it was already a precious place. The area had been designated in 1956 due to the scrabble of rare flowers that had clung onto the high crags, protected from grazing livestock through a combination of precarity and remoteness. This designation proved a bureaucratic obstacle to the return of the wildwood, creating a presumption in favour of maintaining the status quo – the opposite of what the Ashmoles intended for the site. They could only circumvent the 'conservation prison', as Philip refers to it, by promising not to plant trees where they might compete with the protected flora.

In places without these supposed protections, conservationists have more freedom to experiment. The past, however, rarely features in their plans. While the notion of returning the landscape to a previous state is embedded in the concept of ecological restoration, the emphasis has always been more linguistic than practical. If an historical baseline is targeted at all, it is generally in the vaguest possible way: as a muse or poetic device, rather than a set of scientific guidelines. Many conservationists now eschew any reference to the past altogether, preferring to focus on how ecosystems might function in the future rather than how they looked in times gone by.

There are often good reasons for taking this forward-looking approach. On a practical level, most conservationists do not have access to pollen records for the sites they are trying to restore. This kind of work is expensive to undertake and not always possible in any case, given that pollen is only preserved in certain conditions. Even the most accurate records cannot paint a precise picture of how a place would have looked at a particular point in time. Written records, meanwhile, are time-consuming to find and often complicated to interpret, requiring some element of guesswork.

But there are also philosophical objections to looking backwards. For instance: where should we plant the yardstick? Nature is always in flux. It always has been. Trees burn to the ground and new species emerge in their wake. Lakes become marshes. Grassland disappears beneath bramble and briar. The climate warms and cools, causing animals to expand their ranges and then retreat into refugia. The natural world has never experienced one perfect moment of Eden: it has always been a series of novel ecosystems – random assemblages of plants, animals and soils that have arisen in response to unique combinations of climate, geology and disturbance. It is impossible to say what a particular patch of land would look like today minus the impact of humans. Only one thing is certain: it would not look the same as it did 6,000 years ago. The idea of a natural baseline disintegrates the moment you stop to look at it.

More to the point, the conditions we are witnessing today are unlike anything that has occurred in the past. Concentrations of carbon dioxide in the atmosphere are higher than they have been for millions of years. The climate is hot and getting hotter. Soils have been irrevocably altered through centuries of cultivation. Conservationists are rightly asking whether there is any point in recreating the ecosystems of the past if they are unlikely to survive on the planet of the future.

But the division between past and future – between natural and unnatural – is not as binary as it might seem. In his influential writing on woodlands in the seventies and eighties, the historical ecologist George Peterken dismantled the idea that conservationists can aim for either an authentic or an artificial state, instead laying out a spectrum of 'natural' conditions.

When it comes to woodland restoration, he wrote, there are three options. The first is to restore the forest to its present-natural state – how it would look had humans not become a significant influence, while also taking into account the changes in climate and soil that have taken place since the days of the wildwood. This generally equates to maintaining the species that are locally native and removing those that are not. While this is often the default option for restoration efforts today, Peterken did not consider it the perfect solution. It is, after all, impossible to know how the natural world would actually look without the presence of humans. 'Present-natural woodland may be nothing more than an arbitrary collection of native species which happened to be growing on the site when interest in nature conservation developed,' he wrote. 'It may be quite different from what would have been there if the woods had never been altered by people.'[2]

The second option is to allow woodlands to assume a future-natural state – where human influence is removed entirely and nature is left to govern itself. The vegetation that emerges in this scenario will be well-adapted to the soil and climate of the site in question. However, the hardier species would inevitably come to dominate over the delicate. This forest would likely feature invasive species, like rhododendron, and those planted for timber, like Sitka spruce, alongside the native flora. The slow-spreading species that humans have historically erased are unlikely to make it back on their own. Although humans appear to be in the backseat, the resulting forest would nonetheless be diminished by our past decisions. The future-natural option, wrote Peterken, 'carries a strong tinge of nihilism'.[3]

Which leaves the final option: to restore woodland to its original-natural state – where the forest is returned to a truly primeval condition. To attempt such a feat would be largely hubris, warned Peterken. At the most basic level, it would mean reintroducing tree species and eradicating the non-natives that had cropped up over the past few millennia. To achieve further authenticity, conservationists would have to somehow bring back the lost carnivores and mammals, return butterflies to the open glades and beetles to the deadwood, reverse any changes to the soil, transplant lichen onto living bark and inoculate fallen wood with fungi. Even then, the best efforts could only create a facsimile of the original. 'Put in these terms, the objections are obvious,' he wrote: 'The complete restoration of the original-natural woodland would be very expensive, quite impractical and socially unacceptable. Furthermore, even if all practical difficulties could be overcome and society has the patience to sustain the objective for centuries, there is little justification in targeting the forest of 5,000 years ago, or any other particular moment in the past ... Add to that the difficulties of maintaining natural conditions after restoration has been achieved, and one is inclined to dismiss the objective as damaging, unattainable and meaningless.'[4]

All of which appeared to amount to a fairly damning indictment of plans to restore the wildwood at Carrifran. Peterken even turned up to one of the early conferences held by the Wildwood Group, where he sounded a note of caution about their ambitions, warning that the valley was a different place now than it had been in the past. But the Ashmoles were unfazed. To them, the original-natural approach still seemed the most hopeful option. Carrifran offered enough variety that they believed they could find an appropriate place for nearly all the trees that would have grown there 6,000 years ago. Besides, the pursuit of an impossible dream still seemed preferable to surrendering to a tarnished future.

So, on the first day of the new millennium, the first tree was planted in the soil, and the wildwood began its advance – for the second time – across the valley.

I had arranged to meet Philip and Myrtle on a cold weekend in November, although I had begun to suspect that my efforts were cursed. My first visit was cancelled after extreme flooding cut off the road to their house and a second thanks to a bout of illness. Storm Arwen had done her best to prevent me travelling this time, too. The road from Newcastle to the Scottish Borders was littered with trees that had toppled as easily as bowling pins, and my car veered across the lane with the weight of the remaining wind.

When, the next morning, I opened the curtains to see a landscape bright with snow, I began to wonder if it were written in the stars that I should never see Carrifran Wildwood. While I had brought enough layers to feel comfortable crunching a few miles through a frozen wood, I wasn't sure how Philip, 87 years old at the time, would feel about the expedition. But the evening before we were due to meet, he sent me a message inviting me for dinner with his family. I gratefully accepted, and hoped I would get a better sense of whether our walk was still possible, despite the inclement weather.

The Ashmoles' home is a cosy monument to paving your own path through life. The couple have lived here for around half a century; it was the thirty-first place that they considered during a house-hunt that was quixotically focused exclusively on ruins. Every surface is adorned with the clutter of adventure: a beaver skin on the wall, dried flowers hanging from the beams, Myrtle's art and Philip's photography hanging on the walls, and – the *pièce de résistance* – a life-size replica of the Rotten Bottom bow.

This was not what I had expected. Philip retains the well-spoken cadence of the Oxford academic he once was, and I had half-anticipated a world of fading scholarly glamour – a sort of zoological recreation of *The English Patient*. But I found none of the trappings of old money: just a welcome glow from the fire, a home-cooked pie, and the unbridled enthusiasm of Philip, who was undaunted by the gathering snow.

And so, the next day, we pulled on our hiking boots and headed across the hills to Carrifran, where the wildwood was entering its third decade of growth. The route started as a board-walk through a small section of labelled trees – a showcase of the species that they have planted here over the years – before transitioning into the rougher path that winds through the wilder parts of the glen. We were about an hour into our walk when Philip stopped and pointed to the hill that obscured the horizon. At the top, out of sight, was a knoll with a saucer-shaped dip. 'I'm not a romantic but I find it particularly special,' he said, 'because it's quite obviously the place where the hunter would have lain in wait with deer coming by.'

In comparison to the neighbouring glens, stripped by centuries of overgrazing, Carrifran is a tapestry of wonders. Trees freckle the mountainsides, shrinking to scrub and then fading to heathland as the pressures of life above the treeline take their toll. The oaks stand out amid the snow, their bronzed leaves clinging on in reminder of the autumn just past. Holly has kicked free of festive tradition and grows in pagan bursts of waxy green. Tree trunks flaunt skirts of moss and branches wear beards of lichen. Ivy and honeysuckle ascend from floor to canopy in their ceaseless search for light. At higher elevations, cuttings of bearberry have spread into a shaggy carpet of shrubbery across the scree.

It felt to me like the setting of an old folk song – a bleak one, given the weather, probably involving drowned lovers or exiled soldiers. In the warmer seasons to come, however, the valley would explode with a new kind of magnificence, as the trees and flowers unfurl beneath the pale warmth of the northern sunlight. By this time next year, the trees of Carrifran will have another ring to their trunks, and the landscape will be one cycle closer to the grand old wildwood it once was. Now, however, with just over twenty years under its belt, the impression was one of budding growth rather than primeval forest.

Carrifran not only straddles the boundary between youth and antiquity, but also that between wild and manicured. None of these plants arrived in the valley by accident. The trees have all sprouted from seeds that were gathered by volunteers, an operation that took the collectors to some of the remotest spots across the Scottish Borders and the north of England: into steep ravines, down rocky slopes and through ancient nut woods; to the places where it was possible to believe in an unbroken lineage from the original trees of the wildwood to the ones that stood today. Sheep and feral goats are excluded from the valley by a wire fence encircling the whole site, while, in the absence of wolves, any wandering deer are controlled in the countryside way: by a man with a gun.

For a moment, I felt the tension inherent in this approach. Wildness was paramount, and yet the impression was the result of intense micromanagement; each species sought out, every sapling touched by human hand. The young forest was the outcome of decades of physical labour, painstaking research, and an almost obsessive perfectionism. In many ways, it couldn't have been a grander monument to human intervention.

As we mulched through meltwater and mud, Philip told the stories of the trees around us: the debates that underpinned some of the more marginal species, and how they dealt with the vagaries of an imperfect pollen record – the difficulties of distinguishing between different kinds of willow, for instance, or the underrepresentation of aspen despite its continuing presence in a nearby ravine.

Suddenly, something caught his eye: a branch, pruned by a volunteer to prevent it blocking the path. 'I abhor this,' he said. The problem, it transpired, was that the branch had been severed halfway down, leaving a truncated stump ending in fresh wood, too bright and unnaturally smooth – an obvious sign of artifice. Had the branch been cut right back to the trunk, the wood would have healed over and rendered the wound invisible. 'The whole idea is to create a place that is as wild as we can get it, and as soon as you start having branches like that, you can see human intervention very dramatically. Next time I come with a saw, I'll cut it off myself.'

The emphasis on planting at Carrifran jars with the current emphasis on allowing trees to regenerate naturally: on handing the reins back to nature and letting seeds take root where they fall, brought in by the wind or via the digestive system of a passing bird. On the face of it, this seems the best way to create authentically wild areas; indeed, it is how new forests have established themselves for as long as there have been forests. There are certainly benefits to this approach. For one thing, it is cheap: there is no need to pay foresters, or to buy saplings from nurseries, or to erect the plastic tree guards that so often end up littering the countryside. The woodlands that emerge have a more natural aesthetic than those that have been planted by hand, and the plants that grow will be suited to local conditions and require little in the way of aftercare.[5]

When attempting to establish a new forest beside an existing native woodland, this is often the best option. But at Carrifran, there was an obvious problem with this approach. Here, there are almost no trees available to provide the seed source for the future forest: the surrounding hillsides are bare.[6] Had the Ashmoles simply erected a fence around the glen to keep out the grazers, it is possible that some trees would have arrived by themselves – no planting required. However, such a woodland would only ever be a pale imitation of the one that the hunter prowled all those millennia ago, composed exclusively of species whose seeds travel easily or that are economically valuable and have therefore been allowed to persist. What of the slow, the delicate, the unprofitable plants? Must they be banished from the landscape because they are unable to return?

'The danger, in our modern situation, is that you end up with a depauperate woodland because the things that don't disperse well won't get to you in a normal human timescale,' explained Philip. 'We wanted to have a highly diverse woodland, mimicking what was here in the past, on a reasonable timescale. So we had to intervene in a major way, particularly with trees that can't get here naturally.'

There were inevitable imperfections in the reproduction. For one thing, we were perfectly safe. There were no wolves or bears or lynx, and none of the pockmarks and wallows that would once have been created by wild boar and herds of aurochs galumphing in the mud. The deadwood that would have cluttered the original wildwood had not yet made its return; without that rot, much of the original fungi and beetles remained absent. This was still a forest in the first flush of youth; it had not yet developed the scruffiness and ecological richness that accompanies death and decay. Time will plug some, but not all, of these gaps. In a century or so, the woodland will be wilder and more venerable, but the river will never again witness an aurochs drinking from its shallows. Could a lynx ever wander the hills again – or even a wolf? It is possible, but far from certain.

In other ways, however, Carrifran was already a resounding success. The return of the greenery had proved a summons to the surrounding fauna. Before the project began, only meadow pipits made use of the overgrazed hillsides. Now, peregrines nest in the cliffs and hundreds of willow warblers sing in the trees. A few years ago, a tawny owl was spotted for the first time. Frogs spawn in semi-permanent puddles, and red squirrels are occasionally spotted feasting on hazelnuts. Butterflies glitter through the day and moths waft through the night.

But it was the fortunes of the flowers – those rarities clinging to the crags that gave the site its SSSI status – that have perhaps provided the greatest endorsement of the Ashmoles' work. The flora has not been harmed by the return of the trees. In fact, it has expanded. Removal of the grazers has enabled some unexpected species, including mountain sorrel and roseroot, to escape from their precarious ledges and spread downwards through the valley. Sea campion, usually associated with coastal landscapes, has spread along the river right down into the carpark. 'For centuries, these plants have been seeding into watercourses, germinating along the burn, and being shaved off so soon that they haven't even been identified as montane plants,' said Philip. 'We don't know the natural ecology of so many rare species.'

The common adage is that nature knows best: but what if nature has forgotten? Holding the hills in stasis, it seemed, had been denying these plants the opportunity to realise their full potential. Removing the grazing pressure had not swamped them with vegetation: it had allowed them to stretch the muscles that no one knew they possessed. It is a reminder that rewinding the clock is not about the slavish recreation of the landscape at an arbitrary moment in time, but rather refreshing the memory of the natural world as whole.

I see Carrifran not so much as a faithful replica of a vanished ecosystem, but as an attempt to translate an ancient text. There may always be gaps in the poetry – lines irrevocably lost to time – but that does not detract from the beauty of the endeavour. Without the carnivores, the extinct mammals, the landscape here will only ever tell half a story. That is more than we can read almost everywhere else.

Another day, another Scottish glen.

I was further north this time, a little way above Loch Ness, and much had changed since my wintery walk through Carrifran. The snow had been replaced with light drizzle and the flowers were in full bloom. It was spring in the Highlands.

My husband and I had spent the night in a cheap bed-and-breakfast in Inverness, and were making our way towards Glen Affric. The plan was to drive down the length of the loch until the road ran out by the bothy. Then, we would park up and continue on foot for as long as daylight would allow. This seemed the best way to take in the entirety of the glen – a landscape of two distinct halves. The loch stretches along a diagonal line, stretching from north-east to south-west, before tapering off into river and floodplain. The West Affric estate, towards the lower end of the loch, is owned by the National Trust for Scotland, and the vast majority of the rest by Forestry and Land Scotland. The divided ownership reflects the split in the ecosystem itself: ancient forest fading to empty bog as you travel from east to west. It was this strange transformation that we had come so far to see.

We set off early the next day, after a breakfast involving some of the best porridge I have ever had. Through the rolled-down rectangle of the car window, the landscape unfolded like a flip-book. The drama of the difference was enhanced by how quickly one ecotone moulded into the other. The trees on the eastern flanks of the loch were tightly packed: a ballroom dance of birch and Scots pine, their tops tapering into pinnacles pointing towards the sky. Some of them were dead; time had stripped them to the bone, leaving their bleached trunks lying prone across the forest floor. Others were almost entirely swamped by lichen. The air was heady with scents of resin and sap, and the light leaden after filtering through so many leaves and needles.

As we ventured southwards, however, the woodland began to thin out. Absent their competitors, and uninhibited by time and space, the Scots pines seemed to express themselves more fully. They grew singly or clustered in small groves, their muscular branches grasping, Medusa-like, into the air around them. Each tree was worthy of individual admiration. Displayed in this way, splayed out across the landscape, it was easy to see how this particular tree had become a symbol of Scottish identity. It wasn't just its name – despite the *Scots* prefix, the species is widespread across the northern hemisphere – but how they seemed designed to fit within the foreground of the hills. I felt as though I were finally meeting the professors after a lifetime among their students.

Still we continued to drive, and the landscape continued to dilute. The pines became less frequent, and then they were gone. By the time we parked up at the bothy, it felt like we had crossed into another country altogether. The dew of the forest had been replaced by the squelch of bog and heath. There were almost no trees at all; those that had somehow clung on – a few rowans in the gullies and by the waterfalls – were windblown and stunted. Rain concentrated into tiny rivulets that seeped from tussocks of moss and heather. The floodplain, fertile from the regular replenishment of sediment, stretched between the bare hillsides.

The wonders of this side of the glen were more isolated and somehow more ominous. A solitary cuckoo fluted from the other side of the valley. Carnivorous circlets of sundew grew in the acidic soil, waiting for flies. The curled spine of a dead deer, snaked through the grass, discs of cartilage still fresh between the fleshless vertebrae.

A visitor might regard this side of the glen as everything that is wrong with nature in Scotland today: a denuded valley, stripped of the majestic pinewoods that cloak the hillsides only a few miles up the road; a monument to human folly. That was certainly how the National Trust for Scotland perceived the situation when they acquired the West Affric Estate in the early 1990s. Back then, the assumption was that the pinewoods of the east would once have spread through the entire glen, and that humans had been responsible for their demise. Arrangements were made to start restoring the lost native woodland. 'The aspiration was to have this Caledonian pinewood stretching from the east all the way to the west,' recalled Willie Fraser, the long-time ranger for the estate, when I spoke to him on the phone.

A ten-year forestry plan set out priorities including the collection of seeds, on the assumption that planting would take place quickly. Over just a few years, the initial goal of planting 50 hectares of new woodland ballooned to 170.[7] The past was to be brought back to life.

There is an aphorism – often repeated and rarely questioned – that just 1 per cent of the Great Wood of Caledon remains today. This is based upon a figure, published by the Forestry Commission in 1998, showing that pinewoods now cover just 17,900 hectares of land, down from 1.5 million hectares in prehistoric times. The statistic is so embedded in the national consciousness that it was officially recognised by a motion in the Scottish Parliament in 2018.[8]

Rumours of a once-magnificent forest have abounded for millennia. Roman writers, such as Tacitus and Dio Cassius, left behind accounts of a densely forested landscape in which the natives hid from the invading armies as they marched into Scotland. Ptolemy, the second-century mathematician and geographer, was the first to put 'Caledonia Silva' onto a map. But it was the historian and philosopher Hector Boece who really catapulted the concept into legend. His history of Scotland, written in Latin and published in 1527, was the first to directly reference the 'gret wod of Calidon', as his Scots translator would later render it. References to this mythical ecosystem accelerated from the eighteenth century onwards, cropping up everywhere from government reports to academic papers.[9]

It is impossible to piece together a precise picture of the Great Wood of Caledon based upon these texts. Descriptions vary widely by source. However, there are three recurrent themes.

One, the forest was predominantly pinewood. Two, it survived into recent history. And three, its demise was down to humans. A common refrain was that the Romans themselves had cut it down in their attempts to subdue the natives. Others variously blamed the Vikings, English conquerors, ship-builders, iron-smelters and sheep farmers. Blackened stumps of trees, found buried in peat bogs, corroborated the somewhat vague written sources; the fact that they often turned out to be Scots pine further heightened the standing of the tree in the popular imagination. Regardless of the uncertainties around its exact nature and location, the existence of a vast primeval forest seemed beyond doubt, and its destruction nothing less than a disaster.

Over time, the romantic image of the bare Highland hills, so beloved of Victorian travellers, became harder to sustain. The peatlands in which those tree stumps were buried were recast as graveyards and battlegrounds: the wastelands where the Great Wood had once stood, places as unnatural as they were hostile. The Scottish ecologist Frank Fraser Darling memorably described the landscape as a 'wet desert'. The concept was an evocative one, and it stuck. 'In short, the Highlands are a devastated countryside,' he wrote in 1956.[10] Academics and archaeologists added weight to his assertion, blaming prehistoric farmers for initiating the spread of peat through deforestation: removing the trees caused the soils to become waterlogged, they argued, resulting in the build-up of organic material. As one plant scientist put it, in a 1975 study in *Nature*, the spread of bogs across the British Isles was 'a consequence rather than a cause of the demise of forest'.[11]

Forest restoration soon took on a moral dimension. Bringing back the Great Wood of Caledon was an opportunity to make amends for ancestral carelessness – to recover an ecosystem that had been destroyed by greed and violence. There was an element of patriotism to the cause, with the Scottish Green Party among its champions. In 1989, it published a manifesto devoted to the creation of a 'Second' Great Wood of Caledon. The descriptions of the ancient forest echoed those of Fraser Darling, particularly the claim that it had survived almost intact until just over 1,000 years ago. The restoration of this forest, according to the pamphlet, would bring about meaningful jobs, social stability and land reform to rural areas.[12]

There was also a practical element to the task. Interest in the primeval pinewoods was surging around the same time that politicians were panicking about a national timber shortage: supplies of wood had been cut off by German submarines during World War II, and the government wanted to ensure they were prepared for another national emergency. The recent development of wide-tracked tractors and the Cuthbertson plough meant that it was finally possible for foresters to tackle the peaty soils. The Forestry Commission and private individuals, spurred on by generous tax breaks, bought up huge tracts of land for tree-planting – and given that the boggy uplands of Northern England and the Highlands were the cheapest land going, this is where most of the trees ended up.[13]

And so, during the twentieth century, thousands upon thousands of hectares of bogs were drained, ploughed and transformed into forests – if you can call them that. The plantations had the stern geometry of an American city, the trees laid out in tidy blocks, trunks like skyscrapers shooting straight to the sky. It would be difficult to imagine either a Pict or Roman soldier losing their way among these skeletal imposters.

The New Wood of Caledon was a muted place. Draining and ploughing changed the face of the peatlands. Gone were the pools of standing water that once speckled their surface. Rare mosses vanished as the soil shrivelled and cracked. The clear waters of streams and lochs were muddied by sediment, damaging breeding bird habitats and the fisheries upon which the local economies depended. Moorland birds fled and golden eagles sought out new territories as their hunting grounds were swamped by trees.[14] Carbon, locked for millennia in the soils, was released into the atmosphere in huge quantities.

Talk of the Great Wood of Caledon gave a romantic sheen to what was essentially an economic venture. Foresters were not planting Scots pine but rather non-native conifers such as Sitka spruce. That didn't stop the Forestry Commission from speaking of their efforts as a return to the past. 'We are recreating, in difficult circumstances, a small part of that squandered resource; we are restoring a small part of that plundered wealth,' wrote one officer in 1976.[15] Even Fraser Darling was on board. 'The Sitka spruce, and the Cuthbertson plough are a godsend whether you like them or not. They enable us to get cover going in which later we can develop true forest,' he said in a speech to the Commonwealth Forestry Association in 1974.[16]

The trouble was, however, that the Great Wood of Caledon was a myth. Roman soldiers had not encountered an impassable forest as they crossed the northern border. By that point, much of the Scottish landscape had been bare for a very long time.

Of course, the blackened stumps that emerged from the peat hadn't come from nowhere. The trees were real and, in many cases, ancient. It was the cause and timing of their death that had been misrepresented. Again, it was pollen analysis that revealed the real and complex history of the Highland landscape. Research over the last few decades has shown that, following the retreat of the glaciers from Scotland after the last Ice Age, a vast forest did flourish across the breadth of the country. But this wildwood was far from a homogenous blanket of pine. Rather, a mix of species spread across the landscape, governed by variations in climate and soil. There were areas where pine was dominant; where the old myth might have seemed true. But, travelling from north to south around 6,000 years ago, a hunter-gatherer would also have passed between worlds of glittering birch, fruitful hazel and venerable oak, while also perhaps skirting the occasional bare mountaintop or marsh.

The reign of the forest, however, was short-lived. It would be impossible to convey in a few sentences the precise ebbs and flows in extent over the years, or the way the combination of species shifted along invisible gradients, or how the canopy responded to each eruption of wind and fire. What is clear, however, is that the decline of Scotland's woodland happened early and, to a certain extent, naturally. In particular, around 4,000 years ago, the iconic pinewoods collapsed, leaving only scattered fragments and drowned stumps as a reminder of their former glory.[17] The cause of death was not humans but climate change, which initiated the spread of blanket peat. It was these waterlogged soils, not farmers or invaders or shipbuilders, that were responsible for the demise of the Great Wood of Caledon.[18]

For, as the peat deepened, the pinewood vanished. Bit by bit, the Great Wood was swallowed into the mud. So, while it may be true that today's pinewoods cover just 1 per cent of their prehistoric extent, the baseline is an artificial one, based upon an ecosystem that briefly flourished but was always doomed to fail. One may as well declare a biodiversity crisis on the basis that we have lost 100 per cent of the dinosaurs.

'Think of pine as a blossoming – a major expansion and then a contraction back to where it had always been safe,' said Richard Tipping, a former lecturer at the University of Stirling, when I asked him to explain the phenomenon. His work on pollen underpins much of the current understanding about the history of the Scottish landscape, including both Glen Affric and Carrifran. 'Pine is a really flighty thing. It doesn't cope well competing under normal conditions with deciduous trees, but if you change the environment – the hydrology and the climate – then pine gets a foothold and it expands into very unlikely areas. We have short-lived expansions in places which pretty much never saw a pine tree except for these brief periods. It will turn up and grow for as long as it can, maybe a generation or two, and then it fails.'

Even today, the role of humans versus the climate in causing the spread of blanket peat is far from settled. However, with prehistoric farmers having borne the blame for decades, the compass of scientific opinion is now swinging towards an increase in rainfall caused by a shift in atmospheric circulation as the cause. The fact that the pine decline of 4,000 years ago was synchronous across the whole of Scotland supports that hypothesis. Indeed, a study published in 2016 found that, across the UK and elsewhere, the spread of blanket peat from the early Holocene onwards was a wholly natural occurrence, taking place wherever the weather became suitably wet and cold – something that would have happened regardless of whether humans showed up or not.[19]

None of that is to say that people – recent or prehistoric – had no impact on the forests at all. Scotland did not miraculously remain immune to the onslaught of human activity that shaped the rest of the continent. For thousands of years, farming, grazing, felling and burning have chipped away at Scotland's woodlands and prevented their regeneration. Deforestation was particularly intense in the Lowlands, with Carrifran being just one example of a landscape that owes its sterility to humans. However, the notion that people were entirely culpable for creating the 'wet deserts' of the Highlands is an exaggeration. No doubt a Scotland without humans would contain many more trees, but that knowledge must be balanced with an acceptance that much of it would still be bare. When the first communities of farmers arrived in the uplands, they may well have looked out upon a landscape just as windswept and boggy as those we look out upon today.[20]

Bleak, but not barren. Wet, but no desert.

The line between history and myth is often thin. Where facts are scarce, storytelling and wishful thinking are often all that is left to fill the gaps. With repetition and time, these tales become the accepted narrative. The historical ecologist Oliver Rackham railed insistently against the infiltration of myths and pseudohistory into the literature on the landscape, taking aim at the 'grey-minded pseudo-historians' who perpetuate false knowledge. 'It is all the history that most of the public – and some Cabinet Ministers – ever read; much of what passes for conservation is based upon it,' he bemoaned in one particular excoriating diatribe.[21] Similarly, the woodland historian Mairi J. Stewart has criticised the use of history as 'a token used to tart up promotional literature ... or simply used to rationalise pre-conceived goals' rather than as a tool to challenge and enhance conservation work on the ground.[22]

In the worst-case scenario, such as the establishment of timber plantations on deep peat in the Highlands, the misuse of history can have a devastating impact on the natural world. The intention at Glen Affric was never to plant trees for timber, but rather to return legendary pinewoods to their rightful place after so many years of absence – a far more laudable goal. As my husband and I completed our journey on foot through the glen, however, it became patently obvious that the original ambition of the National Trust had not come to pass. The hillsides were stark in their treelessness and the river meandered unimpeded by scrub through the flat expanse of the floodplain. There had been no resurrection of the Great Wood of Caledon. What had caused the change in direction?

Soon after they had purchased the estate, the National Trust took the unorthodox decision to commission a study into the palaeoecology of the glen. The aims were to understand the composition and structure of the ancient woodland and to recommend species for planting upon that basis.[23] The work was undertaken by Richard Tipping and a doctoral student called

Althea Davies, based upon pollen recovered from small rock basins across the site. What they discovered challenged everything that the organisation thought it knew about the estate's natural history and – more controversially – undermined all of the plans for its future.

It turned out that the pinewoods of the east had never spread permanently into the western part of the glen. While the landscape had been wooded from around 10,500 years ago, it was deciduous trees that dominated: a combination of birch, hazel, willow, rowan, aspen and juniper. The proportions of each shifted in response to short-term fluctuations in the weather, but the ecosystem on the whole proved remarkably stable for thousands of years, surviving even alongside the spread of peat. Scots pine made the odd incursion into the glen when conditions allowed, but never stayed long; it is the stumps of those temporary visitors that can still be seen today.

Then, around 4,000 years ago, there was a sudden rupture in the canvas. The weather turned again, with conditions becoming wetter and more oceanic, and the trees disappeared. Not just pine this time, but all the trees. The reason for this collapse remains something of a mystery: the climate at this juncture was not more extreme than the previous shifts. One suggestion was that increased storminess and wind strength might have been the final straw for trees that had already been forced to adapt to life within the peat. Whatever the cause, two things were clear: pinewoods had never been common on this side of the glen, and the demise of the woodland had been a natural affair. The story of West Affric, for thousands of years, was not one of burning forests and hungry sheep, but of thick mud and sphagnum moss.[24]

Suddenly, the establishment of pinewood could no longer be regarded as the return of a lost ecosystem, but the creation of something that had never really settled there in the first place. The collapse of the woodland due to climatic rather than human factors raised the question of whether the National Trust should even be planting trees in the first place.[25]

'The questions that the survey raised made us wary of just blanket bombing the whole area with trees. We cut back on the amount of planting we were going to do. The exclosures were smaller and we put more birch and tree species that were suited to the wetter climate that we had at West Affric,' remembered ranger Willie, who still works on the estate more than two decades later. 'There was some pine planted before the survey, but where we have seen success is more with the broadleaves, like birch. Areas where we planted the Scots pine, although they are growing, they struggle more.'

The difficulty of establishing pine was vindication of the National Trust's decision to base their management of the site upon the back of a pollen study. The choice to keep the glen treeless had, at times, been controversial, leading to accusations that the charity was intentionally maintaining a 'wet desert' at the expense of biodiversity in general.[26] West Affric, after all, was not a museum but a living, breathing ecosystem. The stunted trees that grew in the gullies were proof that trees could be part of the modern landscape, irrespective of what had been there in the past. The tree-planting question was also complicated by the fact that humans first arrived in the glen around the same time as the woodland had collapsed. While it seems likely that these communities were taking advantage of the newly opened landscape, rather than the cause of the change, it is possible that the woodland would have later returned had the land remained unpeopled. The excess of deer, which the National Trust can do little to control, means that self-seeded plants have little chance of growing in the unfenced areas of the glen today.

However, the fact remains that the climate in West Affric today is more like that of 4,000 years ago, when the forest collapsed, than 6,800 years ago, when the pines were at their height. During our walk through the glen, I had to stop on several occasions to wring out my socks; my feet became pale and wrinkled from so much time spent wading through bogs. The weather here has always been wet: now it's getting wetter. Climate change is intensifying the conditions that caused the forests to collapse in the first place. The woodlands that grew during the drier periods of the past may have been more beautiful, perhaps even more biodiverse, but would they be able to cope with the conditions of the future?

'Palaeoecology is a kind of natural experiment: it can help us think through what might happen over decades or centuries in the future, and to understand the interaction between the climate and tree decline and regeneration,' said Althea Davies, now a lecturer at the University of St Andrews, when I asked her to reflect on her work. 'We have never got enough money for all the conservation we want; if we can find the places where woodland is likely to be more resilient in the future, maybe those are the places we should invest in, rather than places where it seems more vulnerable.'

At the same time, historical information must be balanced against contemporary values. The National Trust is not the only organisation with a stake in the future of West Affric. For several decades, the charity Trees for Life has also been active in the glen, planting trees and erecting deer exclosures with the permission of the Trust. Its aim has always been the restoration of woodland, regardless of what the pollen records show, on the basis that the forest would naturally regenerate now if it were not for the sky-high number of deer. Years of hard work are now bearing fruit. Bursts of foliage, surrounded by fences, break up the bleakness of the hills: proof that woodland can thrive here – for now, at least – regardless of the weather and soil. And, if trees can grow here, then why shouldn't they? Why adhere to a particular historical baseline if a better state seems possible?

In the early years, Trees for Life focused, like so many others, on the notion that it was restoring the Great Wood of Caledon. Today, it remains wedded to reforestation, but the emphasis has shifted from the notion of righting an historical wrong towards the general benefits of trees. These benefits, however, are not always unalloyed: studies of the exclosures at Glen Affric show that, while the young forests have attracted a richer assemblage of birds, the disturbance of the peaty soils has released carbon into the atmosphere. Rather than tackling climate change, the trees are currently contributing to it, although that may change as they mature.[27] At a time of competing crises, the answers are never simple. History can provide some direction, but it cannot untangle the mess of the present.[28]

When it comes to choosing what ecosystem to restore, Christopher Smout, who is Scotland's Historiographer Royal and an emeritus professor at the University of St Andrews, says it best. He has probably done more than anyone to bust the myth of the Great Wood of Caledon in the public eye – but that does not mean that he has dismissed the restoration of pinewoods as a worthy ambition. The arguments for this ecosystem, he points out, are rooted not just in ecology or aesthetics, but in our emotions: 'We should be planting new pine woods or extending existing ones not out of nostalgia for some dubious myth, or because we fancy that we owe the past reparation for earlier destruction, but because we are lovers of Scotland and of Scots pine, modern improvers, who choose to treasure the pinewood ecosystem and relish the sight and smell of the woods today,' he writes. 'That should be enough.'[29]

I am not sure what the best outcome is for a place like West Affric. While the National Trust has decided to heed the lessons of the past, Trees for Life is taking a punt on the future. It is not a matter of right and wrong, but of values, priorities and tolerance for risk. Personally, I would rather walk through deep forest and ancient pines than across peat bog and treeless hills – but nature cares little for my aesthetic preferences. Perhaps the trees that Trees for Life have planted will thrive for the next 1,000 years and the valley will bloom again. Or perhaps they will die, and peat will reclaim the glen's contested crown, proving the whole endeavour a waste of time and money. Only one outcome is certain: none of us will be alive for long enough to know.

The story of Carrifran, at least, is a simpler one: a case where history, biodiversity, climate and aesthetics all conveniently align. I returned to the valley again, just a few months after my first visit. I wanted to see what it looked like without the snow, but also I just felt drawn to it. Trudging through the young forest on a cold February day, I had a similar sensation to standing in the centre of an ancient stone circle or leaning back against an old churchyard yew. It was the feeling of sharing a secret across time; of opening a line of communication with people whose faces I would never see and whose lives I could never truly understand. Some 6,000 years separated me from the hunter and his longbow. Within the forest, those years were nothing.

FOUR

Holocene Farm

In the nineteenth century, Victorian travellers developed a taste for Transylvania. *Trans Sylva*. Literally: the land beyond the forest.

At that time, Eastern Europe was still undiscovered territory, at least as far as the British were concerned: somewhere savage and authentic, far from the well-trodden trails of France and Italy; a dark corner of the map where it was still possible to get completely and utterly lost.[1] Transylvania, secluded among the glittering peaks of the Carpathian mountains, was the best that the continent had to offer. Traditional tourism was increasingly viewed with disdain by the upper classes, making the absence of guidebooks a bonus.[2] By confronting such indignities as a soiled tablecloth, a crumbling stove or a loose door-lock – among the observations of one writer – travellers could claim to be bona fide explorers once again.[3]

Transylvanian tourism began in earnest a few decades before the publication of *Dracula* put it on the map for the public at large. People flocked to witness the survival of an older way of life that had all but vanished from industrialised Britain. The region represented a doorway to a lost idyll: here was a world of ancient trees and wild bears, shepherds and gypsies, wooden churches and magic potions.

One such traveller was Emily Gerard, a Scottish writer who spent two years living in Romania in the 1880s, after her Polish husband was appointed to the command of the cavalry brigade. In her memoir, *The Land Beyond the Forest*, she documented her horseback excursions into the mountains.[4] A fearless traveller, she was determined to overcome the limitations that society placed on her gender, bemoaning on one occasion the need to find two male chaperones to accompany the trips undertaken by her and a female companion, just in case the two women should choose 'the self-same moment for swooning away'.

Gerard is mostly remembered for her influence upon Bram Stoker, who incorporated her documentation of local folklore into *Dracula*, having never visited the region himself. But her account is worth reading in its own right: a fascinating snapshot of a world little changed since medieval times, whose natural beauty often had the author in rhapsodies.

'You may ride for hours in the shade of gnarled oak-trees, or, emerging on to an open glade, indulge in a long-stretched gallop over the velvety sward,' she wrote. 'In spring-time these grassy stretches are crowded thick with scented violets, whose purple heads are crushed by dozens at each stride of your horse; and in autumn, when the grass is close cropped, these meadows become one vast playing-ground for legions of brown field-mice, scampering away from under the horse's feet, or peeping at us with beady black eyes from out the porticos of their sheltering holes.'

The landscape that Gerard witnessed from the saddle was far from a pristine wilderness, but rather a rare example of a farming system that had been practised for generations, the foundations of which had been laid down millennia ago: the spiral frieze of Trajan's Column, erected to commemorate the victory of the Romans against Dacia (which overlaps with modern Romania), depicts both forests and haystacks.[5] Even today, there remains a patchwork of meadows, orchards, cropland and wood-pasture that would be familiar to the Romanians, Hungarians and Saxons that lived and farmed this soil so many centuries ago: a landscape that has retained its medieval character despite the pressures of the modern world.

Of course, there have been ripples on the surface. The River Tisza, once said to hold more fish than water, has been suffocated by pollution. A couple of decades ago, fishermen shovelled their corpses from the water – 'copper-coloured like a woman's necklace' – after a gold mine leaked plumes of cyanide into its channels.[6] Climate change has caused outbreaks of pests in the thick Carpathian forests and mountain meadows are disappearing beneath the rising treeline.[7]

The political upheavals have been even more brutal. Romania has weathered two World Wars, the rise of communism and the subsequent return to capitalism. Miraculously, the old ways have survived the tumult. Although rural land was seized by the Soviets to create collective farms, it was eventually returned to its former owners following the fall of the Iron Curtain. Many recipients were happy to abandon the cities and resume life on their tiny plots, raising livestock and making hay as their families had always done. Small-scale farming returned to the Transylvanian countryside just as its last vestiges were vanishing from the west.[8]

Today, the farmers of Transylvania mainly practise subsistence and semi-subsistence agriculture, producing food for themselves and the local market. The land here is far from fully tamed: forests still occupy the ridges and gullies of the mountains, waiting to absorb the smallholdings should the humans ever leave. In the spaces in between, people produce milk, honey, fruit, meat and vegetables. Animals are an essential part of the system. During the summer, sheep and cattle are grazed on mountainside pastures dotted with ancient trees. In winter, they are kept plump on a diet of hay that has been cut from the late-summer fields and piled into haystacks. Woven through these haystacks are the intangible threads of magic, knowledge and tradition, upon which society as a whole depends.

It has long been said that the Inuit have fifty words for snow. Linguists debate whether there is any truth to this claim; nonetheless, it neatly encapsulates how the physical environment leaves its stamp on the language.[9] The concept certainly holds true in Romania. In some villages, there are more than twenty words for hay.[10] There is hazel hay, forest hay and alder hay; soft, rough, dry hay; spiny hay, wiry hay, difficult to swallow hay. There is bad quality hay from marshy areas and the best hay, *fân trifoios*, which is sweetened with clovers. Certain trees – hazel, alder and sycamore – are left standing because of the luxuriant grass that grows beneath their canopies, while birch and oak are known to hinder growth and are therefore removed. Romanian cows are surprisingly discerning customers.

Haymaking begins when 'the bells start ringing in the grass', as the locals put it – that is, when the blooms of the yellow rattle dry up, producing a distinctive tinkling sound as one walks through the meadows, boots knocking at the seeds. Work begins early in the day, when the grass is still wet with dew, or in the cool of the evening. The hay is cut using scythes or electric mowers, turned with pitchforks till dry, and built into haystacks balanced upon racks of dead wood. The tops of the stacks are twisted into crowns to prevent rainwater or snow from seeping inside and causing the spread of mould. Farmers are attuned to the signs of coming rain, like the cry of the owl or the halo that appears around the moon. Seeds that fall from the racks of dried grass are collected and returned to the meadows, ensuring the wildflowers return year after year.

But the calendar of work is not only determined by yellow rattle and the weather: the whims of the saints also help to set the agenda. Grasses that grow at certain altitudes will be tackled around certain feast days; mowing will be entirely forbidden on others. Ignoring these rules can have severe consequences. Each offended saint has their preferred method of punishment. Saint Mary Magdalene and Saint Phocas take revenge with fire and lightning, while Saint Elijah exacts retribution with strong winds. The relationship between the haymaker and the natural world is both practical and profound: a balance of belief, ecology and economy. Ethnographers have coined a word for it: haylife.[11]

There is an abundance of life on these farms that, in itself, seems like a relic from an older world. Brown bears emerge from the forests into wood-pastures, where they feast on ant larvae, fruit and berries.[12] Ancient trees, hollowed and flaking, provide habitats for woodpeckers, beetles and lichen.[13] Butterflies gleam in the pastures and corncrakes call from the meadows.[14] However, what Transylvania is best known for is its flora. The process of haymaking itself creates the perfect conditions for the growth of rare and specialised wildflowers, which are scattered through the meadows like a snapped string of jewels. May sees the arrival of the first orchids – military, green-winged, three-toothed – alongside waves of creamy cowslips. June brings further colour to the landscape, with the addition of dropwort, sainfoin, meadow clary and lady's bedstraw. July adds creeping bellflower and spiked speedwell to the scene. In August, the fields are feathered with the white frills of wild carrot. The performance concludes in September with a display of autumn crocus.[15]

In Transylvania, and places like it, farmers are not enemies of nature. These flowers depend upon them: without their scythes and livestock, the rarities that set this region apart would be swamped by scrubby green. To understand this entanglement of humans and nature, we must travel back in time: back to the very start of the Holocene, when the tendrils of the wildwood were just about to start spreading across Europe's open steppes.

Transylvania sits at the far edge of the Eurasian Steppe – a vast grassy plain stretching around 8,000 kilometres from the Danube Delta in the west to the tip of Russia in the east. Since the end of the last Ice Age, forests have come and gone across this region, advancing in the warmer periods when the climate favoured their growth and retreating when the weather was cool and dry. In the Carpathian Basin, however, there are a few places that appear to have remained unforested since the very start of the Holocene – perhaps even longer. Transylvania is one of them.

Many explanations have been put forward for the long-term openness of the region. Scientists have suggested that warm and dry conditions, including frequent wildfires and prolonged summer droughts, could have prevented the forest from completely overrunning the Transylvanian Plain, even as it dominated other parts of the steppes. Grazing herbivores may also have helped to keep the forest at bay. Although the megafauna had largely vanished by the start of the Holocene, the Carpathians were still home to a number of large mammals, including elk, bison, fallow deer and wild horses, that were capable of exerting significant pressure on the ecosystem. The sunny, south-facing slopes of the mountains, meanwhile, could well have provided a refuge for hard-pressed flora during the last glacial period, while pockets of poor soil, naturally incapable of supporting trees, could have acted as their saviour when the weather grew wetter and the forest advanced around them.[16]

Whatever the reason, time enriched the Transylvanian grasslands. Undisturbed for millennia, a collection of unique and genetically diverse flora was able to develop across the region. When the forest receded elsewhere, it was from this refuge that the new grasslands were repopulated, flowers spilling out from their enclaves into the neighbouring areas.[17]

By the middle of the Holocene, however, even this region was at crisis point. After so many years of stability, the forest seemed poised for victory. As the climate became wetter, the water table rose and fires became less frequent. Without the megafauna to keep them in check, and with the remaining grazers in decline, oak and hazel began to spread across the grassland, increasingly steadily till about 2700 BCE.[18] The trees squeezed the flora into ever smaller pockets, threatening to wipe out the botanical treasures that had accumulated there over the years.

However, before the wildwood could close over completely, a new creature arrived on the scene: the farmer. Small communities settled in the remaining patches of unforested land, where little effort was required to cultivate the soil. Over time, the steppe was transformed into fields and pasture. Their descendants would extend their reach into the forest, burning down the trees and restoring the open land. Sometime during the Iron Age, around 600 BCE, came the invention of the scythe and the creation of the first meadows. Haylife began.[19]

Within this reclaimed grassland, the flowers returned. The farmers, in other words, were performing the same role as herbivores and wildfire had done in the past. Their hunter-gatherer predecessors may have laid the groundwork for the spread of the forest when they wiped out the megafauna, but the rise of agriculture helped to compensate for those mistakes. Through sheep and scythe, humans forged a world that acted as an analogue to the one in which the flora had first evolved. And biodiversity rocketed. Pollen records from Transylvania show a steady increase in floral diversity from the late Neolithic onwards, with the biggest leaps occurring at times of increased human activity: the late Bronze Age, the arrival of the Romans and the onset of the Middle Ages. The only occasions when biodiversity dipped again were when farming was suppressed, after the collapse of the Roman Empire and during the period of cold weather known as the Little Ice Age.[20]

Across the rest of the continent, a similar story was unfolding. Where humans went, flowers followed. In Britain, researchers have found a 'clear increase' in botanical diversity following the introduction of agriculture. In France, richness increased where Iron Age people established cult sites and cemeteries. In the Swiss Alps, diversity 'increased constantly' from around 5050 BCE, when farmers began to burn through the forests. In Sweden, the establishment of fields, wood-pasture and hay meadows prevented boreal forests from erasing the remnants of late-glacial flora, with the only dips in diversity occurring at times of agrarian crisis.[21]

In the twenty-first century, this is not a narrative that we are used to hearing. Farmers are normally cast as the villains of biodiversity, not its saviours. Ecologically, however, it makes sense. In 1978, the American ecologist Joseph H. Connell came up with the Intermediate Disturbance Hypothesis: the idea that biodiversity is greatest when disturbances are neither too rare nor too frequent. Plotted on a graph, it looks like a humpback curve: species richness increases as the landscape fragments in response to disruption and then descends as the intensifying impacts become an impediment to survival.

The disturbance can be anything – wildfires, pests and so on – but farming in particular provides a perfect example of this hypothesis playing out. Small-scale agriculture breaks up the environment, providing a wealth of habitats in a small space: light and shade, hedges and edges, grassland and ancient trees.

Within just a few acres, any number of specialist species might find their niche, while the variety prevents any single one from becoming too dominant. As agriculture intensifies, however, this patchwork landscape becomes increasingly homogenous. Drenched in chemicals, stripped of trees and hedgerows, and devoted to the maximum yield of a single crop, modern farms tend to provide space for only the hardiest of creatures. Pigeons and dandelions thrive at the expense of corncrake and orchid. These days, we are well over the hump of that curve.

It is worth noting that the increase in biodiversity recognised through pollen analysis refers to plantlife alone. But this represents just a small part of the natural world. The first farmers were not conservationists – any rise in floral richness was a happy side effect of meat and crop production. For other elements of the natural world, the rise of agriculture was not something to be welcomed. Initially, scientists believed that the decline of the 'Pleistocene survivors' – the big animals that had avoided extinction during the last Ice Age – began in earnest around 3,000 years ago, due to habitat loss and hunting, as farming started to intensify.[22] However, recent findings suggest that the casualties may have started even earlier – almost as soon as the first farmers arrived on British soil.

Mudflats at Formby, off the west coast of England, brim with ancient footprints. These lie buried beneath sediment until they are exhumed by the waves, allowing a brief glimpse into the ecology of what would once have been a vibrant saltmarsh. For seven years, archaeologists at the University of Manchester made monthly visits to the coastline, recording the prints as they emerged. What they found was a story of startling decline.

During the Mesolithic period, the mud churned with life. The saltmarsh was visited by not only hunter-gatherers, but also aurochs, wolves, lynx, beavers, cranes and wild boar. At one point, the bed was so trampled by red deer that it was challenging to identify any individual hoof-prints among the mess. After the introduction of agriculture, however, the picture changed. Exposed beds from around 5,000 years ago – around one millennium after the arrival of the first farmers – showed only a smattering of red deer, some oystercatchers and small seabirds, and a wolf that may have been a dog. Already, it seems, humans were winning in the battle for space.[23]

As farms ate into the forests, woodland species obviously suffered. In England, beetles that depended upon rotten wood and old birds' nests were rapidly ousted by farmland specialists.[24] Some creatures may only have survived by adapting to life beyond the trees.[25] Even the animals of open land declined. While farmland may have imitated their ancient habitat, it wasn't enough to compensate for the increased conflict with humans. Fossil records suggest that several mammals associated with steppe habitats – including the steppe pika, narrow-headed vole, bobak marmot, lion, leopard, wild horse and Asiatic wild ass – all vanished from the Carpathian Basin around the same time, some 4,000 years ago, as human impacts intensified.[26]

Compared to the losses of the past century or so, however, the damages wrought by early farmers start to look trivial. The decline in biodiversity that has taken place since our grandparents were born has been nothing less than devastating. The palaeoecologist Anthony D. Barnosky writes in his book, *Dodging Extinction*, that these casualties should also be seen in light of the megafaunal extinctions.[27] The earth can only support a certain volume of animals. This carrying capacity is determined by how much energy from the sun is transformed by photosynthesis into consumable sugars and starches. On land, this adds up to 728 exajoules of energy per year – enough to support around 200 million tons of megafaunal biomass, which during the Pleistocene was divided between some 350 species. This figure remained constant for hundreds of thousands of years, before dropping dramatically around the time of the extinctions.

Since then, biomass has been slowly building back up, finally reaching pre-crash levels around 300 years ago, at the time of the Industrial Revolution. This time, however, it wasn't the mammoths and the aurochs and so on that made up most of that weight: it was humans and their livestock.[28] From this point onwards, populations continued to grow. Megafaunal biomass is now eight times larger than it was during the Pleistocene, currently standing at around one-and-a-half billion tons. Fossil fuels enabled us to exceed the earth's natural carrying capacity. Without the option of relying on coal, however, wild animals must continue to depend upon the energy generated by photosynthesis. The trouble is that humans now consume anywhere between 29 and 40 per cent of this energy, too, leaving a much smaller slice for everyone else. 'No wonder other species are starting to go extinct,' writes Barnosky.

Ultimately, a place is more than a tally of its species. Biodiversity is not the only measure of a healthy landscape; it is not even a good one. Plants, animals and pathogens have long crossed oceans and borders through unnatural means, carried on the shoe of a traveller or inside the hull of a ship. While this can boost the number of species in a certain place, it can have catastrophic consequences for native ecosystems, which are ill-adapted to the newcomers. One study found that invasive predators have been implicated in the extinctions of 87 birds, 45 mammals and 10 reptiles worldwide.[29] This game of musical chairs has been underway for thousands of years – we saw, in chapter one, how hunter-gatherers introduced wild boar to Ireland during the Mesolithic period – but it intensified with the spread of agriculture, which provided perfect conditions for plants and insects to spread and thrive. Many of the flowers brought over by farmers in prehistoric times, known as archaeophytes, are now considered a valuable component of Britain's flora: cornflowers and poppies, for instance.[30] Other additions have been less popular. The beetle grain pests introduced by the Romans were probably as welcome to the native Britons as the invading soldiers themselves.[31]

Nature is not a problem that can be solved on a spreadsheet; it is a balance that must be constantly negotiated, based upon what we need from our environment and the debts we owe its non-human inhabitants. Small-scale farming represents an uneasy truce: the land produces food, and humans, in turn, create the disturbances that certain species need to survive, approximating the patchy primeval landscapes in which they thrived for so long.

Transylvania is one of the few places in Europe where this bargain has persisted into the twenty-first century. Whether it will survive into the next, however, is far from certain. The continuation of subsistence farming has largely been an accident of geography: the landholdings here are too scattered, too remote and too marginal to attract the giant farming corporations that now dominate the fertile lowlands elsewhere across Romania. The paper trails of land ownership are often too labyrinthine to trace, making it all but impossible to consolidate the small parcels by buying out individual farmers.

But the long shadow of intensification is beginning to fall across even the most inaccessible corners of the continent. Peasant farmers, trapped in a world without shortcuts, struggle to compete with the cheap offerings of the supermarkets. The elderly are leaving the land in droves and young people are unwilling to take their place, leading to the widespread abandonment of rural areas. This is taking place at such a rate that, by 2030, it is estimated that more than 5 million hectares of agricultural land will have been deserted across the European Union and the United Kingdom.[32]

This is not only a problem for people: it is a problem for nature. Without livestock or farmers, the trees come back quickly. After thousands of years of stability, the steppe-like ecosystem is in danger of disappearing. One study found that, when Transylvanian grasslands were left to become forest, biodiversity decreased among not only plants but also moths and butterflies, and that rare and threatened species were replaced by more common ones. The damage was even more severe within pine plantations.[33] A review of the impacts of land abandonment across Europe more widely found that more than a hundred species would be negatively affected, including eagle owl, capercaillie and Iberian lynx.[34]

Even so, land abandonment is a subject that evades easy generalisation. Some regard the exodus from marginal farmland as a positive step for society: a case of people choosing to swap a life of rural poverty for the promise and opportunity of the city, along with the improved education and healthcare that it brings. From an ecological perspective, conservationists have pointed out that the desertion of farmland does not have to equate to the unfettered return of the forest, and that rewilding – that is, the reinstatement of natural processes through the reintroduction of lost fauna – can help to return the land to something more like its original state.

In 2005, U.S. scientists led by the Cornell ecologist Josh Donlan took this concept to its logical conclusion. In a headline-grabbing paper, they called for the introduction of lions, cheetahs, elephants and camels to North America – an approach they called 'Pleistocene Rewilding'. These animals would act as proxies for lost megafauna, they argued, and catalyse the ecological cascades that had stalled with their demise.[35] Unsurprisingly, the concept was controversial. Other scientists compared the plan to Jurassic Park. The public was outraged. 'Those of us who actually work for a living think you are a colossal asshat,' wrote one complainant.[36] Only in Siberia has the concept really taken flight. In a fenced-off area of tundra, some 4,000 miles east of Moscow, scientists have created Pleistocene Park – an attempt to recreate the lost mammoth steppe by reintroducing various replacements for its original inhabitants. Kalmykian cows, Bactrian camels and musk oxen are among the mammals that have been returned to date. There are no mammoths yet, although advances in gene editing mean that the resurrection of these beasts is no longer such a distant dream. Siberia would be the obvious place to put them.

Mostly, however, rewilding efforts have been more understated. The narrative has shifted towards reinstating natural processes rather than returning the landscape to a particular point in time. Though the principle of reintroduction remains front and centre, the focus has fallen on more socially acceptable species: usually cattle, sometimes horses and pigs, but rarely carnivores. Though less majestic than their prehistoric predecessors, their ecological function remains similar: by trampling and wallowing their way through the landscape, these animals help to unlock the dynamic mosaic of habitats that would have existed in older and wilder times: not just forest and grassland, but also scrub and wetland and wood-pasture. Indeed, the same studies that highlighted the potential losses of land abandonment also found that there were myriad ecological advantages to letting the farmland rough up a little, particularly when this happened at scale, just so long as the entire landscape did not revert to blanket forest. A varied environment means that more creatures are able to find their niche: birds like woodpeckers and treecreepers, for instance, benefit from an increase in trees, while even steppe species like the little bustard can profit from the return of longer and denser grass.[37]

It is humans that stand to suffer the greatest losses from rewilding – at least, when it comes at the expense of small-scale farmland whose abandonment has been forced by economic need. Farmers are the one creature to lose their niche in this environment. While other jobs may emerge, in tourism or conservation for instance, those won't cater for the elderly peasant whose life has been devoted to haymaking. Ultimately, there is no golden state of nature. Agriculture, even when practised on a small scale, involves sacrifice: a human-dominated landscape will always be inimical to wild nature. But so too does rewilding: the restoration of natural processes will inevitably conflict with the cultures and livelihoods that have so long been tied up with the land. Both offer sanctuary, but to different beasts.

In any case, the choice to pursue either approach rarely depends on environmental factors alone. The question comes down to economics, policy and preference. It depends on whether it is still financially viable for farmers to remain on their holdings and whether the government is willing to subsidise them; on whether there is a conservation organisation willing to step in when they leave – or if large corporations, with intensive agriculture or timber plantations on their minds, can outbid them for the land. It depends, too, on cultural memory: how much society values the scent of golden grass, the taste of raw honey, the sight of scythes in the field. On how much we still fear the saints.

Policymakers and conservationists have weighed up these factors and generally drawn the same conclusion: that the demise of Europe's remaining hay meadows would be an own goal for both nature and society. The question is: how best to save them? To find out, I travelled to a place where the medieval way of farming has survived despite the steepest odds of all. To South Wales – the landscape of my own childhood.

The Vile clings onto the edge of the Gower Peninsula. The peculiar name derives from the Old English *gefilde*, meaning field or plain.[38] It is a farm like none I have ever seen.

The fields are lined up like strips of carpet, together leading to the edge of the cliff that drops into the sea. Each one is tiny, around one or two acres in size. From the sky, they look like airport runways, although I recognise how strange this comparison would have seemed to those who tended them for most of their existence. That is because the Vile is special: a working example of how much of Britain would have been farmed during the Middle Ages.

Farmers have most likely been trying to tame this promontory since before the Norman conquest. They cannot have had easy lives. The limestone cliff is bounded on three sides by the ocean. Salty winds whip off the Bristol Channel, bowing the blackthorn hedges and thwarting any vegetables growing too close to the unsheltered edge. The fields have retained their old names, speaking to a long history of struggle against the soil. Stoneyland. Sandyland. Bramble Bush. One, Hawkin Hole, references the mother kestrel that still nests in the rocks, preparing to welcome the next generation of a family as old as many aristocratic lines. The earth sometimes yields tantalising hints of its prior occupants, including, on one occasion, a silver love token.

It was late May when I visited, and the wet weather had delayed the springtime blooms. Instead, the fields were still covered with a weave of weeds, which thickened over the mounds of soil, known as baulks, that divided one strip from the next. With time and warmth, linseed and sweet clover would paint the landscape with stripes of bright yellow and cotton-blue, recreating a scene that had occurred here for many of the last thousand summers. On the edge of the promontory were the hay meadows, almost ready to burst with pollen and petals.

The Vile is a rare example of the open-field system: a method of communal agriculture practised everywhere across Europe until enclosure pushed the peasants off the land. Under this system, each farmer attended his own strip of land, with the members of the village coming together more widely to cooperate and plan in the name of a healthy harvest. Remnants of such farms survive as shadows and undulations across the countryside even today, showing the paths of ox-drawn ploughs as they moved up and down the fields, pushing the soil to the side as they went.

Needless to say, the ancient landscape of Wales would have been totally different to that of Transylvania. The flora of the steppe would have stood no chance against the westerlies whipping off the Atlantic. In prehistoric times, this coastline would have witnessed a shifting cast of ecosystems – forest, fenland, reed-swamp – each of which was ultimately swallowed by sea.[39] However, the ecological principle remains the same: by taking on the role of megafauna and preventing the domination of closed-canopy forest, farmers carved out space for open-land species that would otherwise have been lost to the darkness. In the nooks and crannies of medieval farms, like the Vile, a wide range of plants and animals would have found the conditions they needed to survive. Ground-nesting birds could find cover and camouflage in the fields left fallow – something that was done every few years to allow the soil to recover. Baulks offered safe passage to small mammals as they navigated the cultivated land.[40] The naturalist Colin Tubbs, in a survey of Hampshire, found that only a third of the county's birds were adapted to woodland, with the rest preferring open, marsh, coastal or riverine habitats.[41] Farmers 'inherited the flora and fauna of the more ancient habitats, and indeed, in modifying the landscapes from which they derived, they may have increased plant and animal diversity,' he wrote.[42]

I have been visiting the Gower Peninsula since I was a child; my parents got engaged on Worm's Head, the tapering headland that the poet Dylan Thomas once called 'the very promontory of depression'. I associate this landscape with days at the beach followed by sundaes at Joe's ice-cream parlour in Swansea. By the time I visited the Vile, however, it had been years since my last trip. My sense of nostalgia was cemented by the fact that I had persuaded my mother to drive me there. Sitting in the passenger seat, I felt about ten years old again. As we wound towards the coast, I attempted to refocus my internal vision so that I was peering not 20 but 200 years into the past.

Gower is a place remote from almost everywhere. Like Transylvania, it has been cushioned from the impacts of modernity by the challenges of its geography. The road that leads to the cliffs is narrow, bounded by stone walls and hedgerows. It would not easily admit the kind of heavy machinery that has pounded the rest of the countryside into submission. To approach from any other direction, you would need a boat and a bungee rope.

Even so, the Vile did not emerge completely unscathed from Britain's post-war drive for agricultural efficiency. Certain strips were amalgamated into larger fields, the baulks between them flattened to make way for more crops and larger equipment. When the National Trust purchased the land in 1970, it was well on the way to looking like a modern farm. For years after that, a dual system remained in place: some tenants worked to extract the maximum yield from the expanded fields while others practised gentler husbandry on their medieval strips. Around 2013, however, those tenancies started to come to an end, leaving the National Trust with a decision to make: either renew the leases and continue with the status quo or take over the land themselves and start to shake things up. They chose the latter.

So an unusual experiment got underway. The Trust decided to restore the Vile to how it would have looked during medieval times. With the help of an archaeologist, they began to put back the pieces of the open-field system. Maps, written records and field surveys revealed how the landscape would have been laid out before the war. Volunteers rebuilt the baulks, removed the contaminated topsoil and reseeded the earth, creating the conditions for flora and fauna to flourish once again. Farming continued but with renewed purpose: the yield they were pursuing this time was not crops but nature.

Walking between the strips on that drizzly day, the fruits of their efforts scuttled and scurried before me. A weasel darted across my path and whitethroats sang from the thistles. Hen harriers had returned to hunt in the fields left fallow and the predatory larvae of oil beetles lay in wait for bees upon the flowers. Common fumitory and mustard had sprung up from the disturbed soils. In the meadows, plants like knapweed, speedwell and eyebright were preparing to bloom. 'It doesn't take much to reverse things from where we are,' said Mark Hipkin, the ranger who was showing me around.

It felt like the shallowest layer of a lost inheritance. The deeper strata – the wildwood, the saltmarsh, the reed-swamps – I had missed by a long way. But this felt within reach: a world of abundance that existed within living memory. Had I only been born a little sooner, I might have known it for real.

The Vile is an interesting experiment, but no one – not even the National Trust – is claiming that the revival of medieval agriculture is the best way to feed the nation. There are tens of millions more mouths to feed now than during the Middle Ages. The return of the open-field system would only swap biodiversity loss for mass starvation. That doesn't mean that there are no lessons to be learned from these strange weather-beaten strips: many of the techniques used here, such as leaving uncultivated corridors for animals to cross the fields, can be applied to more intensively managed farmland.

However, as a model for the wider restoration of small-scale farming, the greatest drawback of the Vile is that it is essentially a nature conservation project. The land is not farmed to produce calories or turn a profit, but to support declining wildlife. Half the crop is left standing as food for the birds, and the experiment is partly funded by surfers and day-trippers who pay for the National Trust car park. This is in stark contrast to the function of such farms in centuries past, when productivity was a question of life or death, or in Transylvania today, where the produce grown provides the staples for the community. Research now shows that it is this kind of practical entanglement with the land that produces the greatest returns for nature.

Until fairly recently, hay meadows were a familiar feature of the British landscape. During the 1700s, there were meadows in

Islington, Paddington and St Pancras – places where, nowadays, one would struggle to find so much as a buttercup. Agricultural handbooks from the time reveal the rich traditions that arose around the care of these fields. Football games and bull-baiting events were organised to destroy moss, drive away moles and trample seeds into the soil. Fertiliser could be anything from coal ash to sheep sweat, depending on local availability. Anthills were cherished, to the chagrin of the improvers, on the grounds that they produced an earlier flush of grass and sheltered lambs in the spring.[43]

Today, almost all of Britain's hay meadows have vanished. Modern livestock are fattened on ryegrass that has been pickled into silage, rendering redundant the golden crop of yesteryear. Ryegrass grows thickly, smothering most other species, and matures quickly, allowing farmers to take a cut before any survivors have set seed. This has led to widespread botanical decline. With the average farmer no longer relying on hay to see their animals through the winter, the care of the meadows that remain has either fallen to ecologically-minded throwbacks, who are willing to maintain those in their possession in exchange for government subsidies, or to conservation charities like the National Trust. In both cases, the overriding purpose is not to make hay but rather to sustain floral diversity and the memory of a fading way of life. The story is one that has been repeated across the vast majority of Europe's agricultural land.

Cut adrift from their intended purpose, meadows have also lost their cultural significance. The rituals of haymaking – once governed by a complex litany of seasons, weather, saints, labour, football games, bull-baiting and so on – are now determined by a list of inflexible and inscrutable rules. Farmers who want to claim subsidies under the EU's Common Agricultural Policy (CAP) are subject to strict requirements on when to mow, the number of trees that can be left standing per hectare and the type of machinery they may use to cut the grass.[44] That is not to entirely dismiss this approach: without it, many meadows would have surely been ploughed into the ground entirely. Even so, studies suggest that meadows managed for nature conservation are ultimately poorer in species than those tended to produce winter fodder. The old ways, in this case, are still the best ways.[45]

Walk through the meadows of Transylvania today and you may encounter farmers acting in ways that appear harmful to the natural world. Perhaps you will find a peasant dislodging a carpet of moss with an iron fork or flattening an anthill with a hoe, both being an impediment to hay production.[46] Continue down the road, however, and you could well stumble across a bear gorging on the larvae of an anthill that another landowner decided to leave standing that year. The landscape is the product of thousands of individual decisions, together creating the effect of a patchwork quilt, where almost every species can find what it needs to thrive. It is the freedom to act in their own best interests that has enabled the farmers – ergo this quilt – to persist here so long. If the land no longer served a useful purpose, many would no doubt hang up their scythes, with forests and agribusinesses soon springing up in their wake.

If saints are better than subsidies, then it would seem to make sense to leave the saints to it. Unfortunately, that has not been what's happened. A system that may have worked in preserving the vulnerable meadows of Western Europe has now been transported wholesale to the east. Romania acceded to the European Union in 2007, giving its farmers access to agri-environment payments under the CAP – but only if they agreed to abandon their old traditions and adopt a centralised set of rules. Ironically, it is now these subsidies – designed to enhance biodiversity – that now pose one of the greatest threats to the country's wildlife.

Take the corncrake. This strange, rasping bird thrives in open landscapes: grassy peat bogs, wet marshes and – thanks to humans – hay meadows. The tall vegetation provides cover for their eggs and chicks during the summer. In Britain today, the species is vanishingly rare. Just a century or so ago, however, they were known to nest in people's gardens and keep them awake at night with their incessant calling. The corncrake has only survived in remote places like the Outer Hebrides, where farming has remained small-scale and of low-intensity, allowing their young to survive the harvest. Crofters are incentivised by the Scottish government to delay mowing until August so that the long grass remains standing until the breeding season is complete. It is a measure that has been somewhat successful, in that the bird has not yet gone nationally extinct. Numbers, however, remain low.[47]

Transylvania, on the other hand, is stuffed full of corncrakes. Numbers run into the tens of thousands. This is despite the fact that farmers do not leave mowing until late in the season. Instead, they cut when it suits them: when the conditions are favourable and the saints give the word. They rarely wait till August, by which point the grass has become less nutritious and too tall to mow. The corncrakes, however, do not seem to object.

Again, the patchwork effect comes into play: there is always enough grass for them to find cover somewhere within the quilt.

In 2015, the Romanian government proposed a new scheme to 'protect' the birds. It decided that, to become eligible for EU subsidies, farmers in Transylvania's corncrake hotspots must also refrain from mowing until August. Such a delay would render the hay nutritionally worthless, particularly as climate change now causes the grass to ripen earlier in the year. Unable to feed their animals through the winter, farmers are more likely to abandon their land, thus hastening the demise of the system upon which the corncrake has come to depend. Meanwhile, the rush to mow come the appointed day means that the grass is removed in one fell swoop, which is bad news for any birds still waiting on their nests. In a sudden frenzy of scythes and blades, the squawking, squeaking, rasping miscellany of the meadows is reduced to a homogenous sward. The hurry also means that farmers are no longer able to help one another, breaking down the communal element of haymaking. Without many hands to cut, dry and stack the grasses, the work takes longer, increasing the possibility that the whole lot will get soaked and ruined by rain.[48] And why? To save a bird that does not need saving.

To date, no one has collected the data on how corncrakes have responded to the measure. However, it is not the only species to have faced difficulties in the face of such western imports. In the stretch of the Carpathians that crosses into the Czech Republic, the Danube clouded yellow, a threatened butterfly, was wiped out from its stronghold after the introduction of a regimented mowing regime in the once diverse meadows – an initiative that was once again spurred on by the introduction of EU subsidies.[49]

Farmers accept these subsidies, and the constraints that come with them, because they are poor. In many ways, it is poverty that has been the precondition for the biodiversity of the region – but restricting development, and denying its people the opportunities of greater wealth, is clearly an unethical way to carry out conservation. Indeed, with city jobs and western incomes now a realistic alternative for many young farmers, it is more important than ever to find the financial incentives to enable those who want to to stay on the land. Without these, the spectre of land abandonment will loom ever larger.

Pumping subsidies into rural regions is the simple solution. However, the side effect has been to transform the complex bond between farmers and the natural world into a more transactional relationship. By paying peasants to carry out their traditional roles, policymakers have unwittingly interrupted the complex fugue that plays on in places like Transylvania – that quiet but insistent interweaving of humans, nature and culture.

'Many visiting UK conservationists, especially, fail to accept the need for change in rural Transylvania and instead seek to preserve an idealised countryside that reminds them of an older, lost England,' writes John Akeroyd, an English botanist who has studied Transylvania's flora for some twenty years, in the *Bulletin of the Transilvania University of Braşov*.[50] But Romania is not England, and nor is it lost. If conservationists value the country's grasslands, and the abundance of life they contain, then they must learn to value its people, too.

Akeroyd is the grasslands expert for Fundaţia ADEPT, a charity focusing on the communities and environment of Transylvania. Established in 2004 by Nathaniel Page, a former diplomat at the British Embassy in Bucharest, its work is founded on the ethos that traditional farmers are the best managers of the land; therefore, that the best way to preserve the region's biodiversity is to find ways to make their timeworn practices sustainable within the context of the twenty-first century. 'What we are looking for is not the conservationist's dream,' said Page, when I spoke to him over the phone. 'The hay meadows in Transylvania are this lovely compromise, where they are pretty productive but they are very species rich as well. They are practical landscapes: they are not museum pieces.'

While other conservation charities supported the corncrake measure, ADEPT campaigned against it, successfully reducing the area of grassland that it covered. Rather than chasing subsidies, the organisation focuses on rural development, finding ways for locals to generate a better income from the work they have always done, thereby enabling them to stay where they have always been. In recent years, the organisation has helped to re-establish the traditional pottery centre that had closed during the 1970s, set up a milk processing unit to produce specialty cheese, built a network of mountain bike trails and designed unified branding for local food products. Working at the level of villages and communities, these initiatives are intended to bring money into the region and connect makers to the market – and, ultimately, to allow farmers to remain in the fields.

You will not find, in the meadows and pasture of Transylvania, a blueprint for feeding the world. Nor will you encounter pristine wilderness. But you will find memories: of a way of eating that once fuelled the medieval world; of the primeval steppe, whose richness is preserved in the fields where that food is produced. You will find meat and milk and honey; orchids and butterflies and bears. It is a world in balance – a balance that is becoming increasingly disturbed as the outside world creeps in. There can be no locking the door when modernity comes knocking. But the farmers of this place must be empowered to turn the key on their own terms. If only to keep the corncrake. If only to appease the saints.

FIVE

Our Place

On the lake lived a family of swans. They floated silently, pearlescent against the perfect blue of the water, keeping a wary distance from our little boat: hunters kept their canoes here, too. We had come not for the waterfowl, however, but the fish. I was in North Karelia, Finland, where I was beginning to learn that a life close to nature is necessarily a life close to death.

In the boat beside me were two fishers, Karoliina and Lauri. We had motored a short way from shore, before shutting off the engine and using the paddles to row between a series of traps set out across the lake, their positions marked out by flags. It was late October, and soon winter would spread across the landscape. But in these last gasps of autumn, at least, the world hummed with energy. The morning light was golden as sap, the shoreline draped in evergreen and birch.

Karoliina, not wearing gloves, plunged her hand into the water and pulled out a hooped net, spangled with fish. She untangled each one gently, careful not to damage the ropes, before passing them to Lauri, who dealt a blow to the brain and a crunching stab through the gills. The plastic bucket at my feet began to fill with blood and silver, stray scales floating to the surface like froth. Some determined individuals leapt back to life after minutes or more of seeming dead, requiring a second flick from Lauri's blade. Fishing this way was a slow and deliberate process; a significant investment of time per mouthful of flesh.

Away from the artificial light of the supermarket fish counter, the finery of each species became more sharply defined. Most common was bream, followed by perch with its spiny dorsal fin. On one occasion, Karoliina pulled out a burbot, glamorous in its mottled leopard-skin coat. 'It never feels like work,' she told me, her hands pink with cold. 'Always like an adventure.'

Most Finns, however, would be unimpressed with this haul. Bream, perch and burbot are all considered 'trash fish'. They are flat and bony and difficult to prepare; the less delicious cousins of salmon and trout. The catch that Karoliina and Lauri had painstakingly pulled from the water was destined to become fish mince or pet food. The prime fillets that would end up on people's dinner plates, meanwhile, would be sourced from away fish farms.

'In the old days, people used to pay a high price for bream. Now they are too much in a hurry,' said Lauri. He was solemn and serious-minded; the kind of person who, when driving us home later that day, took care to point out the ancient pines that had survived alongside the busy road. 'Much easier to have Norway salmon with cream, salt and pepper. Thirty minutes and dinner is ready.'

With all the nets examined, we hopped in the van and headed down the road to the fish base to process the catch. Karoliina and Lauri slapped the bodies onto the wooden table and sliced off the pink-white fillets, making it look easy. When I attempted it myself, I was surprised at the coarseness of the uncooked flesh, how tightly it clung to the delicate bones. I asked Karoliina how she learned so quickly. She was a city-dweller, still a student, and this was her first year on the job. 'Videos on YouTube,' she replied cheerfully.

In a couple of months' time, open-water fishing of this kind would become impossible. Finland's many thousands of lakes would freeze over, like so many shards of shattered glass, and the number of daylight hours would dwindle. But that did not mean that Karoliina and Lauri would stop fishing: only that the task would become colder, darker, more physically intense. Their boat would be swapped for snowmobiles and they would turn to seine nets to catch their prey.

Seine nets are enormous – hundreds of metres long – and are used to scoop up rather than entangle the fish. The nets encircle the shoals, which are heaved to the surface at the end of the day. The method has been around for thousands of years, carried out in the open waters of spring and summertime everywhere from New Zealand to Egypt. However, Finland is one of the few places where seining is pursued throughout the winter, when the lakes have closed over with ice. For centuries, perhaps longer, villagers have come together to harvest fish in seemingly impossible conditions. On Lake Puruvesi, where Karoliina and Lauri spend the season, seiners target vendace, a small fish that swims in shoals through dimly lit water, commonly fried in rye flour and butter and served as finger-food, eaten bones and all.

Until the mid-twentieth century, winter seining was carried out manually. Horses hauled the equipment into position, holes were cut into the ice by hand, and the nets were guided through the water using long wooden poles. Fishing in this way was a communal task: the demands and dangers of the icescape meant it had to be. But the necessity of collaboration also meant that seining acted as something of a social security system: everyone who fished was entitled to a portion of the catch.[1]

On Lake Puruvesi, modern technology has made the task a little less gruelling, if somewhat less romantic. The horses were replaced by snowmobiles in the 1970s, and the wooden poles by radio-operated ice torpedoes around a decade later. When I visited, operations were run from a purpose-built warehouse owned by the local municipality alongside the lake. The fish were processed on stainless steel countertops and vacuum-packed in plastic. Such innovations haven't been entirely positive: the equipment runs on petrol, making it more expensive and polluting to run than the animals of old, while increased efficiency means that seining is no longer such a communal endeavour. But they have, at least, enabled the fishery to survive. When Karoliina and Lauri step onto the ice this winter, they will be continuing a gnarled and secretive tradition that is fading fast from Finland's lakes.

Globalisation, urbanisation and climate change have left little space for an activity that depends upon cold weather and rural cohesion. Where the seiners once fished for subsistence alone, today they are participants in the market economy, earning an income by selling their catch to wholesalers who sell it on to supermarkets and chains. They have struggled to compete with the cheap produce of fish farms and trawlers, where the environment, rather than the consumer, is forced to pay the difference. Attitudes have also changed: for the younger generation, the bright lights and warm offices of the city offer an appealing alternative to the sub-zero conditions of the lake. The elders are dying, and with them the knowledge of how to navigate these icescapes. In the community of Rymättylä, in south-west Finland, it was climate change that delivered the final blow to the tradition: the mild winters of the 1990s meant there was practically no ice left to walk on when the time for seining arrived.[2]

The winter seiners of Lake Puruvesi have not been immune to these pressures.[3] Here, there can be no economies of scale; no corners that can be cut without risking the lives of the fishers themselves. Already, as temperatures warm, the ice is forming later and breaking up earlier, cutting the fishing season by up to half and reducing the profit margins that are already squeezed, as well as making it increasingly dangerous to spend time on the ice. Fish are behaving unpredictably, undermining the seiners' knowledge of where to lower their nets. While the price of vendace has been suppressed by the national fish markets, the cost of the fuel required to run the equipment has increased. In the 1960s, almost sixty crews operated on Puruvesi during the winter. Today, there are fewer than ten.

These remaining crews, however, nurture an ember of knowledge – of understanding – that has softly burned for centuries. Winter seining is not just about securing a meal or even an income: it is an alternative way of knowing the world. Out on the ice, fishers are led not only by the physical conditions but by the phases of the moon; older seiners will sometimes dream about the locations where they will find the best catch. The lake is carved up into hundreds of sites known as *apaja*, knowledge of which is passed down orally through the generations, with names that can stretch back into prehistory. The act of seining features in the *Kalevala* – the country's national epic, collated in the nineteenth century based upon the oral poetry sung by rune-singers since time immemorial – demonstrating the prominence of the tradition within Finnish and Karelian cosmology. One of the final songs in the poem tells of how fire was unleashed from the belly of a whitefish, after being caught within a seine net by the old magician Väinämöinen, before sweeping through the forests of spruce and juniper, and ultimately burning 'half the North'.

By keeping the practice of winter seining alive, the fishers of Lake Puruvesi are maintaining the last whispers of a way of life that was once commonplace across Finnish society. No one states it explicitly, but Karoliina and Lauri bear a responsibility that feels impossible to skirt. Unless Finland's youth can be persuaded to re-embrace the ice, the current generation of seiners could well be the last.

Finland arrived late to industrialisation. Around 200 years ago, the country remained among the poorest in the world. Most of the population were farmers, who fought against short growing seasons to produce meagre harvests from small plots of land. Although not entirely sheltered from the rapid development underway elsewhere across the continent, it was World War II and its aftermath that ultimately catapulted the nation into modernity. Finland, which had sided with the Nazis in an attempt to avoid the seemingly greater menace of the Soviet Union, was made to pay massive reparations to the latter in the form of ships, machines and wood products, forcing a total upheaval of its economy.

Sheltered from the outside world, Finland's rural population continued to hold onto its folk magic and traditional knowledge long after such worldviews had vanished elsewhere across Europe. The old ways remained particularly embedded in North Karelia, where the majority of the songs in the *Kalevala* were collected from the rune-singers for whom they were still a living tradition.

The old beliefs of the Finns are too rich, too sprawling, to be contained within a few paragraphs: suffice it to say that nature was not something inanimate and detached, but rather fluid and interconnected. Theirs was a landscape that could be mastered through incantations; where the forest was alive and sacred, inhabited by sylvan spirits known as maidens of the woods, who looked like humans from the front and trees from behind; in which boundaries between objects were thin and shape-shifting was common. Entities had their own supernatural life-force – the untranslatable concept of *väki* – applied to everything from water to rocks to sauna steam.[4] Protective charms were hung upon the walls of the cottages and buried beneath their foundations. Visiting a family in North Karelia in 1901, the folklore collector Jaakko Lonkainen noted, among other things, the presence of a thunderstone, the tooth of a bear, the head of a viper, the whiskers of a black cat and a tinderbox inherited from an ancestor. 'All of these have their own individual power and they are part of the fortune of the house,' he wrote.[5]

This mystical perception of the world was shaped by encounters with nature: the forests, lakes and marshes in which Finns remained deeply immersed. The tales that they told one another, write the archaeologists Vesa-Pekka Herva and Timo Ylimaunu, 'can be conceived as aids for perceiving the richness of the lived-in world; telling the stories guided listeners to notice such aspects of the environment that would otherwise have been more difficult to perceive or remained hidden'.[6] In other words, the people belonged to the landscape. By spending time in nature, fostering an intimacy with its rhythms and undertones, the deeper layers of the landscape were revealed to them in turn.

With industrialisation, however, came the widespread desecration of the natural world. During the second half of the twentieth century, rural people moved to the cities in droves, surrendering their trees to the sawmills. The primeval forests that once covered the majority of Finland were cut down and turned into timber, paper and pulp. By the twenty-first century, in the southern half of the country where industry was most intense, old-growth comprised just 2 per cent of the standing forests.[7] The rest were managed plantations: young, skeletal, conspicuously tidy – and destined for the chop. Wildlife struggled to maintain its foothold in these cut-and-paste ecosystems. Creatures that rely on deadwood have fared particularly badly: almost all wood is harvested in the effort to maximise profits, leaving little behind for those that depend on death and decay.[8] One study found that 1,000 species – half the country's forest specialists – are already doomed to extinction in the south, as there is not enough habitat remaining to support viable populations.[9]

Nor were forests the only ecosystems to suffer. When suitable soils for establishing plantations ran out, the state forestry company turned to the nation's peatlands, ditching and draining them to support spruce and pine.[10] All that wood needed transporting somehow, and so the rivers were engineered into featureless channels along which the logs could float unimpeded to the sawmills downstream. Lakes turned browner and cloudier as sediment leached from the land and into the water, causing damage to cascade through the food web, affecting everything from fish to waterfowl.[11]

Over time, the mystical worldview that had persisted for so long was ploughed back into the soil, vanishing in tandem with the ecosystems that had nurtured it. For city-dwellers today, the charms and rituals of the pre-war world are more dream than memory. Sylvan spirits exert about as much influence on a resident of Helsinki as leprechauns do upon a Dubliner. Even for conservationists, folklore is regarded as immaterial when it comes to the preservation of the natural world. But Tero Mustonen is not the average conservationist.

My first encounter with Tero took place at 8.15 a.m. in the car park outside my flat in Joensuu, the city where I was staying for the week. He rolled up in his mud-spattered car and I hopped into the backseat. He was headed to the fish base at Lake Puruvesi, along with Karoliina and Lauri, where they had a meeting with a government health inspector. 'Welcome to North

Karelia,' he said, as I settled onto the back seat. 'Or North Korea. I don't see much difference.'

Tero is a man who wears many hats. He is a fisherman, the youngest person who can still fish the icescapes of Puruvesi entirely by hand; an adjunct professor of human geography, who co-led a recent report from the UN's Intergovernmental Panel on Climate Change; the leader of Selkie, the small village where he lives with his wife, Kaisu; and the president of Snowchange Cooperative, an organisation he helped set up in 2000 to protect the environment and traditions of the north. He is deeply embedded in the landscapes and culture of North Karelia – although it is an inheritance he could have easily disavowed. While his grandparents were 'truly voodoo, part of the old culture', his childhood was spent in the city, where he went to school because 'that's what you have to do'. He later joined the army and then worked as a wilderness guide, before going on to study at university. It was only after that he began fishing in earnest, apprenticing himself to three older fishers who taught him the skills he would need to earn a living upon the ice.

I met with Tero several times during my stay in Finland: he was generous with his time, and willing to answer my questions about the land and traditions that he knew so well. Over a few days, he showed me around peatlands, islands and lakes; introduced me to old forests, new forests, burned forests; and gave me lunch at his home, a building some two centuries old, with baskets of ripe fruit in the porch and gloves of moose fur hanging on the wall.

Despite this, I would be lying if I said I ever felt comfortable in his company. Tero was sure-footed in his relationship with the natural world, and I felt myself stumbling mentally as I tried to keep up. He had little patience for the visitors who came seeking quick fixes and cheap sentiment: for the volunteers who hoped to heal themselves without also building connection with nature; for the London tourists who refused to kill a burbot; for environmentalists who condemned the ethics of the hunt. Nature, he stressed often, is not about pretty scenery and simple cures: it is about death and the surrender of self. About putting the collective before the individual. I gave up trying to find the words that might secure his esteem, and instead resigned myself to sitting with the flimsiness of my own connection with the earth. The truth was, I didn't have the words. His world was deep, frozen, bloody, earthy. Mine seemed gossamer by comparison.

It was his work through Snowchange that had really brought me to North Karelia. Back at my computer in England, I had read about how, since its inception, the organisation had put traditional knowledge and community rights at the heart of its environmentalism. Its early projects had focused on documenting the oral knowledge of herders and fishermen and suchlike, on the basis that such information could underpin the restoration of the landscape, both ecologically and spiritually. It had also funded more practical interventions, such as nomadic schools for Indigenous children and the installation of solar panels in reindeer herders' camps.[12] Only later did the organisation expand into more active conservation, with the launch of its Landscape Rewilding programme in 2017. Today, Snowchange is one of the largest landowners in Finland, barring timber companies and the state itself, with thousands of hectares of peatlands, wetlands, forests and river catchments in its care. Even there, people have remained front and centre, with their interventions guided by the traditional knowledge of those who know the area and the rewilded sites co-managed by the local community.

It was an approach that attracted me. The discussions around rewilding I had witnessed at home in Britain seemed marred by vitriol, with those who lived and worked in the countryside often feeling threatened by those who sought a wilder future for the land. I wondered if there was something to be learned from the conversations underway in the north. I hadn't known what to expect when I turned up in Finland, but it certainly wasn't a fish processing warehouse. Rewilding is often portrayed as the removal of human intervention; of allowing nature to take the reins. The growth of a fishery, even a small-scale one, seemed the opposite of that. However, it was through Snowchange's rewilding programme, under Tero's personal guidance, that Karoliina and Lauri were learning to seine, and so it was to Lake Puruvesi that morning that I found myself bound.

After an hour or so in the car, including a stop for coffee and cinnamon buns, we reached the lakeside. The weather was sunny but the wind was strong, and the surface of the water churned tempestuously beneath its fingers. Roped to the shoreline was a shallow canoe, carved by one of the elders the previous spring. The wood was waterproofed with pine tar, its curves tailored to the precise patterns of waves as they rebounded off Puruvesi's many islets. But it mainly existed for the purpose of tourist trips. Tero handed me a life jacket and ushered me into the small motorboat beside it.

As we bumped along the waves, he explained the rationale behind including the Puruvesi fishery among their rewilding initiatives: the seiners, it turns out, were helping to preserve the clarity of the water. Although Puruvesi is still relatively clean, eutrophication has nonetheless been underway since the 1980s. The influx of organic material and nutrients from the catchment area has caused occasional algae blooms in the once-crystal bays – the slimy side effect of decades of industrial exploitation. By removing large quantities of vendace, but not so many as to deplete the population as a whole, the fishers removed excess nutrients and kept the waterscape in balance.[13] 'You will not find a better example of a human-nature coupled ecosystem,' said Tero.

The humans – their knowledge and their stories – weren't some optional extra in the rewilding of Puruvesi. They were fundamental to it. The people here have always needed the wilderness. What I learned from Tero that day was the other side of the story: that the wilderness needed the people, too.

Rewilding is a concept that, in its very name, invokes a return to wilderness. And the wilderness, in the minds of most people, is not populated by fishers on snowmobiles: it is empty, rugged and remote. The idea reached its apotheosis in America with the Wilderness Act of 1964, which defined such lands as places 'untrammelled by man, where man himself is a visitor who does not remain,' and has since infiltrated the minds and strategies of conservationists across the globe.

Over the last century or so, this perception of wilderness has inspired acts of great brutality. In an attempt to preserve nature in its supposedly Edenic state, conservationists established national parks and nature reserves through a model known as 'fortress conservation'. This approach began in North America, with the exclusion of Native Americans from their ancestral lands, and would ultimately prove one of the nation's most successful exports; today, two-thirds of the world's terrestrial protected areas are in the global south.[14] Across the world, Indigenous people were cast as environmental villains to be evicted in the name of saving nature – and they have been, en masse. Between 1990 and 2014, more than 250,000 people were expelled from their land, often violently.[15]

The rewilding movement across Europe has thankfully shunned such egregious violence – but the association with depopulation remains. As we saw in the previous chapter, rewilding is what happens in places like Transylvania when people no longer have a role to play in the landscape. In some cases, it has been rewilding itself that has led to the economic pressures that have forced people to find a living elsewhere: in Scotland, the price of land has surged as so-called 'green lairds' have purchased estates on which to deliver their environmental ambitions, making it harder for communities to regain control over the land where they live and work.[16] Rewilding organisations claim to bring employment to the local area, which may be true in a sense: in 2021, the charity Rewilding Britain published analysis showing that their work had resulted in an increase of around 70 full-time equivalent jobs over a 10-year period.[17] Many of these jobs, however, were in tourism, conservation and education – sectors in which people are observers, or at best stewards, of the environment. As a result, it is often those whose lives are tied most closely to the land who have most resisted its return to nature.

We have already seen, in the previous chapter, how small-scale farming can offer a way for humans to remain on the land in a culturally meaningful way. Farming, however, inevitably demands a transformation of the wild into a landscape centred around human needs. What Tero sought was something different: a way for people to exist within the wilderness, without turning it into anything other than it had always been. For humans to belong to the land again, just as they had done for tens of thousands of years.

I ended the first chapter of this book by asking a question: have humans ever been wild at all? In a paper published in 2018, two academics – an anthropologist and an ecologist – argued that, for most of history, humans should be regarded not as disruptors of the natural order but a part of it: a keystone species whose actions were as integral to the ecosystem as the creatures with which they coexisted.[18] Other plants and animals had evolved alongside human activity – including digging, burning, travel and predation – and had come to depend on it for their own survival. 'Just as with the loss of other keystone species, these co-evolutionary relationships can unravel when such societies are displaced via colonialism, or their interactions substantially and rapidly altered by radically changing political and economic circumstances,' they wrote. To restore today's ecosystems to an authentically natural state, they argued, rewilding must reinstate not only the lost ecological functions of megafauna but also people.

The western concept of wilderness excludes this notion of humans as a keystone species. The Finnish language, however, has its own word for the concept: *erämaa*. It is not an exact translation, but rather describes a landscape in which humans are intermittently present: a place neither entirely civilised nor completely untouched. The word harkens back to a way of life that may have been in place as early as the Stone Age, where villages possessed wide territories through which residents would disperse to hunt and fish.[19] This system broke down from the sixteenth century onwards, when the Swedish king, who ruled Finland at the time, declared such unsettled areas to be the property of the Crown.[20]

While the absorption of the *erämaa* territories by the Crown did not prevent Finns from hunting and fishing in the usual way, it fundamentally changed the relationship between the villagers and their land. Collective rights to the wilderness were overwritten by the rights of the state. Landscapes once woven with stories and knowledge came to be regarded as vacant spaces, ripe for industrial exploitation – and, as Finland's need for natural resources expanded during the twentieth century, exploited they were.[21] For Tero, this was not an abstract problem. When we last spoke, he had just lost a case in the Supreme Court against a mining company that had put in a claim to explore for rare earth metals across an area that covered the whole of his village. Half the catchment of Lake Puruvesi was under a similar claim. While Finland is justly celebrated today for its principle of Everyman's Rights, which allows anyone to fish, forage and camp in the wild, this success is only half the story.

Folklore, community rights and rewilding are often treated as separate realms. But listening to Tero speak, hearing about his on-the-ground experience, it became clear how closely the three were entwined. People without rights were unable to protect the land that birthed their stories. Without their stories, the people became disconnected from the land. Disconnected from people, the ecosystem fell out of balance. By putting humans first, Snowchange was putting nature first, too. There was no distinction.

'What the west, for want of a better word, calls "wilderness" has been the cultural landscape for many of our peoples for thousands of years. The complexity of how the seal hunt happens, or how fishing happens in a winter community, is so profound, so integral to human-nature relationships, and it produces human beings that are very integrated into that landscape. These are the kinds of nuanced ways of living that are hard to demonstrate when it's so polarised: that you either use the lands or you protect them,' said Tero.

'That is why we have tried to advance the notion of a third way, where, for any village, it's simply clear we need to restore our habitats and conserve our forests. But that doesn't mean we can't use some of the resources within the customary law and governance we still have. And that understanding goes far beyond the modern discourse on conservation ... Rewilding is a very powerful mechanism, if you know how to integrate cultural knowledge into it. It becomes a vehicle for equality and empowerment, and community and Indigenous rights. But that doesn't happen often.'

Of course, Finland today is a modern nation-state. The winter seiners of Lake Puruvesi have no time for relics: an immutable tradition is a dead one. No one, not even Tero, was arguing for the revival of a Stone Age lifestyle in North Karelia. His may have been a world of forests, ice and incantations, but it was just as much a world of YouTube videos, climate change and motorboats. Rather than seeking to revive winter seining as it looked a century ago, his mission was to find a way for the tradition to survive into the future. For the fishers to continue to earn a living upon the lake in a meaningful way – not as performers of old Finnish culture for the benefit of passing tourists, but as guardians of the last fragments of an alternative way of life.

The training of Karoliina and Lauri was one component of that mission. But what it would really come down to was whether Snowchange could raise the value of vendace. If the pursuit of 'trash fish' were economically sustainable once more, the fishery had a chance of attracting new recruits: people to whom a life on the ice appealed, but who nonetheless needed to make an income. For more than a decade, Tero and the older fishers of Puruvesi have been raising the reputation of vendace at home and abroad to this end. In 2013, the EU awarded the vendace caught from Lake Puruvesi protected geographical indicator status, in recognition of the traditional way they are caught, as well as their unique silvery scales and soft bones, caused by the unusual clarity of the water.[22] The tradition of winter seining has been inscribed on Finland's National Inventory of Living Heritage, a UNESCO initiative, alongside sauna bathing and Santa Claus.[23] Snowchange has revamped the marketing for vendace, emphasising the role of the fishery in maintaining the health of the lake.

Tero hopes that these efforts will, in time, transform vendace into the delicacy it deserves to be – with the price tag to match. 'Romantic longings for some distant past are useless. We have to live today,' he told me. 'That doesn't invalidate the significance of traditional practices, but that society doesn't exist anymore, and that's a fact we have to embrace.'

What I took home with me from Finland was not an easy message about reintegrating humans into the wilderness. There was no escaping the modern world and the market economy, not even at -30°C in the middle of a frozen lake. The days when the humans here could count on the land to meet their every need were gone. What I did learn, however, is that rewilding is not a zero-sum game: that there is still space for humans in the wild – if we make it. But equally, that it is incumbent upon us to remember how to belong to the natural world. To become creatures that the wild might absorb once more. To reclaim our role as a keystone species.

Back in England, sat behind my desk, the act of remembrance began to weigh more heavily upon me. What were we meant to be remembering, exactly? The British had not been hunter-gatherers for thousands of years. Hunting and wildfowling had been the preserve of the elite since the Anglo-Saxons, while the old remedies and recipes that depended upon wild plants had long been forgotten or fallen out of use.[24] Nor was I naïve enough to think that those who lived in the countryside, now or in the past, offered a flawless model of human-nature relations. Wolves and beavers did not vanish on their own – and there are plenty who would rather they never returned.

Even so, the wild did not disappear from our lives in one fell swoop. The agrarian landscape may have been a largely human creation, but it still contained areas rich in biodiversity – peatlands, wetlands, forests and meadows – that abounded in useful material.

While Anglo-Saxons largely depended on farmed crops and livestock for food, wild resources played a continuing, if fluctuating, role in their diet and beyond. Sloes, plums and elderberries were plucked from hedgerows and woodland edges to make syrups and dyes.[25] Rivers, lakes and coastal waters provided conger eels, mackerel, mussels and limpets. Wetlands were a source of wildfowl, including the ever-popular crane meat. A tenth-century *Herbarium* directs readers to the specific habitats where they might find medicinal plants, indicating that they grew wild rather than in herb gardens tended for that purpose.[26] Wild animals were a source of symbolism, too: amulets of bear claws, eagle talons and beaver teeth have been recovered from the graves of women and children, suggesting a link to fertility or protection.[27] The decoupling of humans and nature may have been well underway by this point, but it was far from complete.

Customary rights to the use of the land were eroded following the Norman conquest, with the imposition of Forest Law – which, despite its name, applied to a variety of non-wooded habitats – over vast swathes of England, designed to protect the king's access to deer. Following widespread civil strife, however, those rights were reinstated in 1217 with the Charter of the Forest. The vellum document is testimony to the myriad ways in which people remained reliant upon the wild, acting as the basis for 'an economy grounded in natural resources', as one professor of law put it.[28] It asserted their rights to graze pigs and cattle among the trees, to collect bracken, harvest bark, gather firewood, cut peat, and scoop up honey from wild beehives. These materials would have been used for meat, thatching, soap, leather, furniture, tools, heat, mead and medicine.

Such customs underpinned peasants' lives for centuries, bringing with them a degree of self-reliance, but the system was not to last. By the 1500s, a land grab was underway again, reducing the area in which people could practise their rights. The perpetrators were motivated not by venison but wool. Sheep, wrote Sir Thomas More in *Utopia*, had become 'so great devourers and so wild, that they eat up, and swallow down the very men themselves'. The semi-wild world that had sustained the medieval populace through lean times and long winters was enclosed behind a network of hedgerows, fences and stone walls. Over time, people became less dependent upon its bounty. As the backlash against privatisation gathered steam in the nineteenth century, the emphasis was on the recreational rather than economic value of the land. The Victorian magazine *Once a Week* gave an account of this 'new claimant' to the country's commons: 'He does not demand special rights of pasture, estover, or turbary;[29] all he asks for is the privilege of wandering about these breezy spots, of playing on them, of breathing the fresh air that sweeps over them.'

Reliance on the wild declined further during the twentieth century, as society grew increasingly wealthy, urban and industrialised. There was a brief resurgence during the 1940s, when rosehips played an unexpected role in the war effort: the government initiated a scheme whereby schoolchildren were paid to collect the small red pods, which were then turned into syrup, providing a vital source of vitamin C as imports of oranges grew scarce.

By the twenty-first century, however, the economy had almost – but not entirely – decoupled from the wild. In 2001, two scientists from the Royal Botanic Gardens at Kew surveyed the commercial uses of wild and traditionally managed plants across England and Scotland. Their research uncovered the small but significant contribution that wild lands continued to make to rural wallets: patches of nettles whose leaves were used to wrap cheese; wetlands that provided reeds for thatching, saw-sedge for weaving and club rush for sealing the oak panels of whisky casks; whinberries gathered from the moors and baked into pies, crumbles and muffins. But the report carried the air of documenting a dying tradition. In the future, the authors noted, their work would either provide a snapshot of 'more lost features of a culturally and biologically richer past, or a springboard for integrating plants, conservation and employment in a way that supports both local economic activity and a sense of local or regional identity'.[30]

In the decade or so since the publication of that report, it seems that the former scenario has come to pass. When it comes to the wild, my own country is among the least connected across the whole of Europe. The UK lags behind only the Netherlands in its aversion to untamed produce, with just 8 per cent of households heading to the forest to forage in 2016, mostly targeting berries.[31] Wild mushrooms, which are collected by the bucketful elsewhere across Europe, are almost entirely shunned. The reasons for such disenchantment are political and cultural – the result of having been alienated from nature for so long – while our mycophobia (fear of fungi) possibly extends back to the Anglo-Saxons, with the word 'toadstool' carrying connotations of poison and lechery.[32] While pharmacists in France are trained to know which mushrooms are safe to eat, the British are generally discouraged from mushroom picking altogether, having grown up with warnings from parents and scare stories in the media.

Of course, I am not advocating for the return of a pre-agrarian society. The last time Britain was inhabited by hunter-gatherers, the island's population stood somewhere between 2,750 and 5,500, meaning that up to 200 square km of land was needed to support a single person.[33] With a population of nearly 70 million today, this would clearly be unsustainable – even if it were desirable. Moreover, in the modern world, there are myriad ways for people to get closer to the wild that do involve getting one's hands dirty. Community-owned renewable energy and improved broadband are among the less romantic solutions for helping people out of the city and into more remote areas; we were not all born to be cooks and craftsmen, after all. On its own, the wild harvest is never going to be a game-changer for rural development. Ultimately, the opportunities for employment will always be limited, by both the availability of produce and the seasons themselves. But none of that means we should discount it altogether.

What wild produce offers, to those willing to accept it, is a means of income diversification, in a way that also adds value to the natural world. Elsewhere in Europe, foraged goods make a meaningful contribution to both the economy and people's lives. The variety of products gathered from the woods reflects an intimacy with their ecology that has been all but lost in Britain. In the basket of a continental forager you might find truffles, bilberries, chanterelles, walnuts, acorns, wild thyme, pine cones, birch sap and conifer resin – products destined not just for the dinner table but also for medicines and decoration. Collecting is most popular in the east: in Latvia, for instance, almost 70 per cent of households will go foraging at some point during the year.

Much of the produce is consumed by the collectors themselves, or given as gifts, but a significant proportion is also sold: for 1.5 per cent of households, these products comprise up to half their income. A recent study found that, across Europe, the value of forest products – excluding wood – adds up to almost €20 billion in total, which averages out at some €78 per hectare of wooded land. Official estimates, which exclude the value of self-consumed products, had put it at a fraction of that.[34] In other words, policymakers are significantly undervaluing our forests. In tending these landscapes for timber alone, managers overlook an opportunity for both nature and people. In the long run, a forest flowing with nuts, berries, fungi and herbs may well be worth more than a stand of desiccated stumps. When it comes to the potential for biodiversity and belonging, such a forest is undoubtedly the more precious.

The cottage industries of Eastern Europe may seem a poor business model for the more economically developed UK. However, while enthusiasm for collecting diminishes as wealth grows, the products gathered in richer countries tend to sell at a higher price. It is actually in Western Europe where the wild harvest provides the greatest boost to the value of the forests: in Switzerland, it added more than €300 per hectare, compared to just €19 in Ukraine. After Russia, the countries where wild produce contributes the most to the economy are France and Germany.[35] This is because collectors in wealthier countries are more able to add value to their products, using processing facilities and marketing strategies to capitalise upon the demand for authenticity and ethical consumption – and there are plenty who are willing to pay.

I had my own taste of this industry over a decade ago, when I spent a part of my final university summer holiday volunteering in Iceland, picking blueberries and mushrooms from heathlands and timber plantations, filling cardboard boxes with produce to be packaged up and sold to the local shops. That month remains one of the most wonderful of my life, when I felt connected to the pulse of the land in a way I had not before, nor have done since.

Back at home, however, the opportunities to embrace my new hobby were slim. Britain is one of the most nature-depleted countries in Europe. Access to wilder spaces, particularly in England, is severely limited, while foraging, even for personal use, is illegal on private land. I was not brought up knowing the names of plants, nor how to distinguish the edible from the poisonous – but, even if I had possessed the knowledge that enabled our Mesolithic ancestors to make full use of the land, I would have lacked the space, abundance and diversity to use it. Species once common have become scarce; habitats once extensive are now small and fragmented. Just last summer I learned that it was possible to eat wood sorrel, a small plant with drooping flowers, associated with ancient woodland. I found a patch on a moss-covered wall while walking in the countryside near my home and popped a heart-shaped leaf onto my tongue. My surprise at its sharp lemony taste was tempered by a pang of doubt: was there enough here for me to eat, or would I do better to leave well alone?

Overharvesting is a concern, of course, particularly if more people are encouraged to dip into nature's larder. In 2022, residents of the Cornish village of Lostwithiel proposed patrols of their verges to protect the wild garlic from outside restaurant suppliers intent upon taking the lot.[36] In Epping Forest, on the outskirts of London, there have been warnings that commercial fungi collectors are damaging the biodiversity of the ancient habitat.[37] Overexploitation becomes a growing risk as particular ingredients become trendy, leading to an influx of collectors to places within an easy distance of towns and cities. More generally, however, studies show that most foragers are concerned with the sustainability of the products they harvest, often taking active steps to ensure they do not deplete their supply.[38] Indeed, Oliver Rackham points out that the woodlands that survived into the present day tended to be those whose economic or social value to the landowner prevented them from being grubbed up for farmland.[39]

The balance may be fragile, and the implications unappealing, but the message remains as true as it has ever been: money is a force for protection. We do not throw diamonds in the dustbin. It is time we learned to value the treasures of the wild in a similar fashion. Doing so would not only preserve existing habitats: it could catalyse the creation of new ones. Overharvesting, in the end, is the result of too many people, targeting too few species, in too little space. Our hunger for the wild carries within it the seeds of opportunity. It provides not only an incentive to rewild our depleted island, but a means to pay for it, too. How much would a hectare of forest – or coastline, wetland, heath, and so on – in Britain be worth if we accounted for the true extent of its wealth? If, instead of turning to carbon credits and tourism trips to fund the restoration of the natural world, we turned instead to blackberries, mussels, hazelnuts, samphire, chanterelles, reeds and birch sap?

That was the essence of the question that I put to Marian Bruce on a hot day in July. Alongside her husband, Simon Montador, she runs Highland Boundary, a distillery that creates spirits and liqueurs from the berries, weeds and foliage that they gather from the land around their home in Perthshire, as well as the neighbouring forest, where they have a licence to collect. The previous autumn, I had given out their produce as gifts at my wedding – bottles of larch and honeysuckle spirit, of birch and sloe liqueur – and I was intrigued to hear more about their story.

Bruce is a molecular epidemiologist by training. In 2008, however, she left her academic career and suburban house in Stirling and moved to a seven-acre patch of overgrazed farmland with the intention of putting back the lost vegetation. Together with Montador, she planted trees and hedgerows, reintroduced the flora of the understorey, and dug a new pond, all while continuing to breed a small flock of Hebridean sheep. Today, the farmland is half-wild and half-cultivated: a microcosm of the dualistic world in which humans thrived for thousands of years. Bruce describes it as a forest garden, akin to the semi-wilderness that the Europeans found when they landed upon the shores of North America for the first time, where favoured foodstuffs were encouraged within the context of a natural landscape.

'It is not a new thing at all. Actually, humans have always done that,' she told me. 'Forest harvesting is quite different because it looks like a forest, rather than a field with a fence around it, and we haven't had that for probably a thousand years in the UK – that's all gone here, so there's not a cultural knowledge of that kind of thing. People have this idea that farms have to be monocultures, but it is much better to have as much biodiversity as possible in what you're going to harvest: it makes your farm much more resilient to disease, but it also allows you to have different income streams.'

On their own farm, rewilding had come before the distillery. But since launching Highland Boundary, Bruce has become something of a thought leader in the realm of wild produce – particularly when it came to incentivising the generation of richer landscapes – working alongside universities, charities, government ministers and fellow foragers to ease the path for others who might want to follow in her footsteps.

'We are building the economic case for rewilding,' she said. 'It's not just about tourism. What we do is an example of the kind of green jobs or nature-positive businesses that can come from a rewilded landscape ... We are under-utilising our forests, and there are many more things we can harvest sustainably from them without chopping them down.'

Bruce is not just in pursuit of ecological revival. At the heart of her mission is her hope for a cultural renaissance: a means of reconnecting people with the plants that would have once filled their recipe books, medicine cabinets and stories. 'As a nation, we are quite interested in exotic and shiny things, and that tradition just doesn't exist – is not passed down from generation to generation – as it is across the rest of Europe,' she said. 'When we started to go out and talk to people about what we do, what became apparent was that it was a really powerful way of connecting people back to the landscape, and to plants, and things they hadn't really thought about or thought were important.'

Her words chimed with everything I had heard from Tero. Here was that same triad of people, culture and ecology, transported from the frozen lakes of Finland and onto the honeysuckle-scented hillsides of Scotland. It struck me once again how just how recently our species has become disentangled from the wild. How recently the wild has become disentangled from us. What Tero and Bruce both seemed to offer was a pattern for re-plaiting those threads: of restoring that interdependence, while at the same time sucking the poison from a debate that has long been more venomous than it needs to be.

Because conservation should have never become about displacing people from the land. The wilderness is not a blank slate: until recently, it was our home, our larder, our library. The trouble is that we have forgotten how to live there. Abandoning our role as a keystone species, we forfeited our rights to its deeds. The wild harvest offers a way for humans to reclaim something of that role; to expand our minds at the same time as expanding the wilderness itself. Seen in this way, rewilding becomes not an act of severance but revival: for earth and humans both.

SIX

Where the Wild Things Were

I am going to tell you three stories. All are about animals; all about loss.

The first takes place in Kiribati, a nation comprising 33 islands stretched out across 3.5 million square km of the Pacific Ocean. Almost all of these islands are atolls: hollowed rings of coral, each fringing a turquoise lagoon, that have formed around the submerged rims of extinct volcanoes. Today, the reefs of the Gilbert Islands, the nation's main archipelago, are frequented by fifteen species of shark. Some are familiar, like tiger, blue and whale sharks. Other species, like the cookiecutter and sicklefin sharks, have names that wouldn't look out of place on the jars of an old-fashioned sweetshop.

Fifteen species of shark may seem like a lot of sharks. For someone used to oceans that contain almost no sharks at all – or for anyone attempting to swim between the atolls – it is a lot of sharks. After more than a century of commercial fishing, however, the reefs around Kiribati are emptier than they have ever been. The species that remain are the remnants of a richer past.

Islanders have always relied upon the ocean for resources, including shark meat. Left to their own devices, fishermen would only take the predators in small numbers. In 1892, however, Kiribati was placed under the protectorate of the British Empire. The colonial administration had an eye to profit, and it wasn't long until they realised the potential of the market for fins. Reports of commercial shark fishing commenced in 1910. After World War II, the Colony Wholesale Society, which controlled the economy, encouraged locals to build their stocks of fins, even running a publicity campaign in anticipation of a booming trade in China. It was not very successful. 'Sharks are cumbersome beasts which cannot easily be hauled into a fragile craft, and have to be towed back to shore once they exceed a certain size,' according to a 1957 report from the South Pacific Commission. The fishermen, it continued, 'neglected to take the fins although knowing that co-operative organisations would buy them'.[1]

Even so, by 1950 around 3,500 kg of shark fins were being exported from across the archipelago every year. Numbers remained at around that level for the next few decades.[2] Commercial fishing only ended in 2016, when the government of Kiribati, independent since 1979, established a sanctuary within its waters.[3]

By that point, however, populations had already plummeted. Sharks are sensitive to overhunting. They grow slowly, mature late, and produce only a small number of young. Hammerheads had disappeared entirely: historical literature shows that the species, currently absent from Kiribati, swam its reefs as recently as the 1930s. The lack of any written records before the twentieth century, however, meant that it was impossible to know the full extent of the loss. There were no accounts of how the reefs had looked before the arrival of the Europeans; no stories that might have provided a glimpse into the pre-colonial ecology of the ocean.

Or so it was believed. Hidden away in a museum archive, however, was a collection of artefacts that hinted at the abundance of former times: objects so unconventional that they had been overlooked as a potential source of natural history. Weapons.

Land and resources have always been at a premium in Kiribati. Historically, feuds were settled through ritualised combat. These weapons were fashioned from one of the few materials widely available in this world of sand and water: shark teeth. Warriors attacked their foes with 18 ft spears, while their henchmen carried tridents. The bodies of these weapons were carved from the wood of coconut palms, with rows of shark teeth lashed to the edges using coconut fibres and human hair. Despite their fearsome appearance, they were generally intended to wound rather than slay. Fighters defended themselves from serious injury with puffer fish helmets and breastplates made from porcupine ray skin.

Following the arrival of the Europeans, shark tooth weapons became an object of intense curiosity: exotic artefacts to be bartered and brought home by the various missionaries, whalers, traders, curators and colonial administrators that passed through the islands. Over time, conflicts subsided, and weapons were no longer the useful possessions they once were. Islanders were happy to trade them in for tools and tobacco. Thus, a steady stream of teeth flowed from battlegrounds of the Pacific and into the glass cabinets of museums across England and America, where they could be ogled by tourists somewhere between the drawers of pinned butterflies and the overpriced gift shops.[4]

It was while working in one such institution – the Field Museum of Natural History in Chicago – that a fish biologist called Joshua Drew came across a collection of these weapons. They had been kept in the anthropology department, but Drew realised that they had ecological value, too. The shape and serration patterns on the teeth were well-preserved, meaning that it was possible to identify the species to which each had belonged. The weapons were mostly manufactured between 1840 and 1898, meaning they pre-dated the colonial era. They were, in other words, a window into the historical reefscapes of Kiribati; a snapshot of the sharks that swam among them before the ransack of the fishing industry began.

Drew and his colleagues identified eight species of shark in total. The most common was the silvertip shark, whose teeth appeared in 34 weapons. But it was the discovery of teeth belonging to dusky and spottail sharks that surprised them: neither had been encountered in the water around the Gilbert Islands for more than a hundred years. Both are absent from the written data collected during the last century. In fact, the weapons were the only evidence they had ever been there in the first place.

The commercial importance of these species means that it is unlikely they were overlooked in surveys and, since there are no records to suggest trade with the Solomon Islands or Fiji – the nearest places where the sharks are present today – it is improbable that they were sourced from elsewhere. The most likely explanation is that the sharks were wiped out by humans.[5] Dusky and spottail sharks, wrote Drew and his colleagues, were an example of 'shadow biodiversity' – creatures that had been lost before the advent of formal scientific assessment.

Creatures that we have come to believe do not belong in the present, because we have forgotten they were there in the past.

The second story takes the form of a poem, *The Song of Heledd*, written in Wales sometime between the eighth and tenth centuries, although it concerns events that happened during the 600s.

Heledd was the sister of Cynddylan, prince of Powys. But he was slain, alongside the rest of her family, in an almighty battle with the English, for which she was somehow responsible. In the poem, Heledd alone remains, guilt-ridden and weeping, to contemplate all that she has lost. She mourns the destruction of her home, dark and roofless where once it was warmed by fire and feasting, and sings of the clovers in the church graveyard that have been reddened with blood. But most striking – most terrifying – is her description of the two eagles, of Eli and Pengwern, that have arrived in the aftermath of battle to gorge on the blood of the dead.

The eagle of Eli watches over the seas:
Fish do not penetrate into the estuaries
He calls, he feasts on the blood of warriors.

The eagle of Eli travels through the woods tonight.
His feasting is to his fill.
The violence of he who indulges him succeeds.[6]

Eagles are a regular trope of medieval poetry. Alongside the wolf and the raven, they are one of the 'beasts of battle' that turn up to feast on carrion following a battle scene; the trio also features in *Beowulf*, having a boastful chat about how many corpses they have eaten. But the poet in this case was unusually specific regarding the diet and habitat of the birds. The eagles that Heledd describes have emptied the estuaries of fish and taken them back to the wood to eat. It is not much to go on, but it is enough for a naturalist to make an identification. This is not some vague avian metaphor: it is a white-tailed eagle.[7]

The historic presence of white-tailed eagles in Wales has long been debated. Fossils provide some evidence that the raptor at least passed through in prehistory. But remains are sparse: the archaeological record contains just four occurrences of the bird, which could have feasibly been transported to their final resting places by human hand.[8] There is little written evidence for the species before the 1880s, and those that do exist are disappointingly vague.[9] Ecologists had concluded that the white-tailed eagle was only ever a rare visitor to Wales, passing through but never stopping to breed.[10] This perception has hampered efforts to reintroduce the bird in modern times, as official guidelines require animals to be returned within their historical range.

The Song of Heledd, however, suggests that the white-tailed eagle was alive and well in Powys until at least the seventh century. Of course, poetry is a shaky foundation for science.

Highly stylised stanzas of unknown provenance are never going to pass peer review; they would certainly be a bad foundation for any high-stakes reintroduction projects. However, this poem is far from the only evidence that eagles, both white-tailed and golden, were once a widespread feature of the skies. The charisma of these birds mean that they have never been just passive residents of a landscape: rather, they defined it. The bards honoured their presence in verse. Ordinary people named places after them.

Maps of Wales are peppered with place-names recalling the one-time presence of eagles. There are more than sixty locations incorporating *eryr*, the Welsh word for the species, with occurrences particularly common in the north. Many of these place-names also include a reference to habitat, which scholars take to mean that the reference is literal rather than metaphorical. It also allows ecologists to differentiate between the two species: white-tailed eagles nest in trees and prefer lowland coastal areas, while golden eagles nest in crags and hunt on higher ground. Pass through the countryside and you will quite possibly stumble across one of these tiny eulogies to the country's lost predators. *Nant-yr-Eryr*, Eagle Stream. *Coed Eryr*, Eagle Wood. *Cefn-Yr-Eryr*, Eagle Ridge.[11]

The eagles didn't stop at the border, and neither did the names. Across the British Isles there are hundreds of such toponymic memorials. There are references to eagles in Gaelic, Manx, Norse, Scots, English and Old English, illustrating the vast expanse of time and space through which eagles once soared. Based upon such evidence, conservationists have calculated that Britain and Ireland, around 500 CE, would have supported up to

1,400 pairs of golden eagles and another 1,500 pairs of white-tailed eagles.[12] By the nineteenth century, however, both species were rarities. By the 1920s, white-tailed eagles had disappeared entirely and golden eagles were hanging on by the skin of their beaks in the Scottish Highlands, victims of human persecution and habitat loss.

'He is gory – I do not defy him,' trembled Heledd of the eagle that fed upon her brother. A millennium later, and such fears had lost their hold; the birds had fallen victim to our own thirst for blood.

Thanks to conservation efforts, both species are now back in Scotland. White-tailed eagles have recently been reintroduced to England, too: the government has issued a licence for the release of sixty birds over a five-year period on the Isle of Wight. Wales, on the other hand, remains eagle free. Although plans are currently afoot to return the white-tailed variety to the south-east and the Severn Estuary, a modern-day Heledd would currently find little to fear in the woods. Given the extent of persecution and habitat loss that raptors have suffered in recent centuries, we are lucky to have eagles in Britain at all. Even so, it is difficult not to mourn the lost abundance of the Dark Age landscape, when these birds were garrisoned on each crag, patrollers of every woodland, watchmen of every stream – at least, if the bards can be believed.

The third story is about John 'Grizzly' Adams. Born in Massachusetts in 1812, he was variously employed as a circus-man, shoemaker, miner and rancher – proving outrageously bad at all of them. He was almost killed by a Bengal tiger; lost his entire stock of shoes in a fire; gave up his mining claims in reckless speculation; and had thousands of dollars' worth of cattle stolen in a single night.

Finally admitting defeat, Grizzly took to the road in 1852, heading into the Sierra Nevada mountain range of California, where he 'resolved thenceforth to make the wilderness my home and wild beasts my companions'. He meant this in the most literal sense, trapping several bears – he called them Lady Washington, Benjamin Franklin, General Fremont and Funny Joe – and keeping them as companions. He later returned to San Francisco and then New York, where he earned a boom-and-bust living with his performing menagerie of wild animals. He died in 1860, aged forty-eight, after an old head wound, inflicted by a wild bear, was re-opened by a monkey that he had been training for the circus.

Grizzly's exploits were recorded by Theodore Hittell,[13] a reporter for San Francisco's *Daily Evening Bulletin*. Following hours of interviews with the man himself, Hittel would ultimately ghostwrite Grizzly's autobiography: *The Adventures of James Capen Adams, Mountaineer and Grizzly Bear Hunter of California*. These memoirs provide an entertaining – if idealised – account of his life. The misfortunes of his youth are skipped over in the first four pages, while the wife and children that he abandoned in New England do not feature at all.

Nature, on the other hand, is writ large throughout the pages: the book is a paean to wilderness, and to the machismo of those who passed through it. Grizzly and his companions ford rivers, climb gorges, are awoken from sleep by the howling of wolves. They feast upon a giddying array of wild animals (including the bears that weren't turned into pets) and drink tea brewed from mountain herbs. Camping out on an island in the Carson River, Grizzly recalls how they caught 'a fine lot of salmon-trout, using grasshoppers for bait, and in the night killed half a dozen beavers, which were very tame'.

It was this throwaway remark that recently caught the attention of ecologists in present-day California. Because there shouldn't have been any beavers to catch at all. Decades of scientific orthodoxy, upheld by the California Department of Fish and Game, held that the rodent was historically absent from large parts of the state. Those that did show up were considered a non-native irritation, harmful to both ecosystems and farmland, and were treated as such. Even today, it is still illegal to relocate a 'problem' beaver: state law instead requires the unfortunate creature be killed, a task that many approach more in the spirit of vengeance than pest control, if the numerous online videos of Americans blowing up beaver dams with explosives are anything to go by.

This perception of the Golden State beavers was based upon the conclusions of two respected American naturalists. In 1937, Joseph Grinnell published *The Fur-Bearing Mammals of California*, in which he concluded that beavers had historically been absent from the Sierra Nevada, the Bay Area, the entire coast and almost all of the south. In 1942, the zoologist Donald Tappe published a monograph, *The Status of Beavers in California*, where he essentially confirmed the theory. 'There are no known records of beavers ever having occurred in the Sierra

Nevada,' he wrote, 'except where these mammals have been recently introduced there by man.'

Which all sounds very decisive, until you notice the footnote. An addendum from Tappe notes quietly that, since writing the above, he had heard from a Mr Roy Mighels, who spent much of his teenage years during the late 1800s riding horses through the mountains. During that time, Mighels observed plenty of evidence for beavers along the Carson River – exactly where Grizzly Adams claimed to have trapped half a dozen around four decades prior. 'He attributes the disappearance of beavers from the Carson River drainage to the heavy trapping done in that area prior to 1900,' wrote Tappe. 'It seems, therefore, that beavers actually did inhabit at least a part of the eastern slope of the Sierra Nevada south of Lassen County.'

In a few lines of small text, consigned to the bottom of the page, Tappe undermined not only his own theory, but also the preconceptions that would inform state policy for the next eighty years. He also stumbled upon the cause of his error: the fur trade.

Before the California Gold Rush, there was the California Fur Rush. For more than a century, trappers had ransacked the nation's rivers for beavers. By the mid-twentieth century, when Grinnell and Tappe were writing, there were simply none left to find: most of them were adorning the heads of fashionistas in Paris and London. Nor were there any physical specimens in the natural history collections of nearby museums. As a result, rather than concluding that the rodents had been extirpated from much of California, the two naturalists imagined that they had never been present at all.

Grinnell and Tappe overlooked the accounts that the trappers had left of their expeditions. Admittedly, there were not many of these available to them: unlike Grizzly Adams, these travellers headed into the mountains looking for profit rather than a story to tell. They often misreported their bounty to hide the best hunting grounds from others. In the last decade, however, researchers have uncovered missives that challenge the perception of a beaverless landscape.[14] A report from 1831, by a British-Canadian trapper called Peter Skene Ogden, celebrated the capture of a thousand beavers from the tributaries of the San Joaquin River, which flows from the Sierra Nevada right into San Francisco Bay.

These trapping expeditions were so successful that it appeared as though the beavers had never been there in the first place. And the lack of bones in California's museums? That was because the museums themselves were too recent. Their collections were all initiated after 1904, by which point the mountains had already been reduced to a beaver desert. To confirm their theory, researchers would have to wait for physical evidence to emerge – which it eventually did. In 1986, an episode of flooding revealed several old beaver dams, high in the mountains, buried beneath more than a metre of mud. The wood was so well preserved that it was possible to make out the incisions where the beavers had gnawed at the tree. Radiocarbon dating confirmed that one of these dams was first constructed around 580 CE, before finally being abandoned around 1850 CE, around the peak of the fur trade. Since then, even more relics of old beaver dams have been exposed in places where the animal supposedly never set foot.[15]

In these ancient dams lay proof of what the mountain men had said all along. The beavers had not been absent. They had been forgotten.

The common theme throughout these three stories should now be obvious. For hundreds of years, humans have obliterated wildlife populations to such an extent that even the memory of them has gone. This loss of memory is one of the greatest hindrances to conservation today: forgotten species, rather than being reintroduced, are often actively resisted, considered dangerous non-natives rather than the vulnerable exiles they really are.

In the tales I have just told, it was only chance encounters with unconventional sources that recalled their presence. Luckily, there is an abundance of such material available to those who care to seek it out. The first ever drawing, daubed on the wall of an Indonesian cave more than 45,000 years ago, shows a trio of warty pigs. The painter of this scene inaugurated a tradition of animal-based recordkeeping that has only intensified with time. Telling stories about wildlife is a very human obsession.

Our fascination with animals is exceeded only by our need for them. We rely on their flesh for food, their fur for clothing, their courage for protection. We see our own identities echoed in their essence: it is why we display eagles on coats of arms, lions on sports kits, bears on flags. In their assumed characteristics we see teachable moments: animals feature strongly in nursery rhymes, fairytales and fables. Their mystique is at the heart of so much of our myth. We fear them, too: humans are not only predators but also prey.

The fanatical nature of our recordkeeping reflects this dependence: it makes sense to know if stocks are declining, if wolves are nearby, or herds on the move. Nowadays, these accounts have found another purpose. They provide a baseline for wildlife before humanity pushed it to the brink. Such information is not easily found in traditional scientific sources, and so it is to dusty archives and museum cupboards – to logbooks, encyclopedias, letters, memoirs, place-names, poetry, weaponry – that conservationists must turn.

Such material can be difficult to interpret. Written accounts are generally not produced by disinterested naturalists, but by hunters, fishers, settlers, poets, politicians and artists – people who were either uninterested in the truth or who sought to conceal it. Their words must be taken with a grain of salt. Colonialists embellished the natural bounty of new lands, hoping to persuade others to join them. Politicians succumbed to ulterior motives. Poachers concealed their crimes. Poets slip easily between folklore and fact. Even the supposedly impartial naturalist was prone to bias: old notebooks are more likely to contain evidence of rare, charismatic and unusual species than of the everyday occurrences.

No wonder, then, that the path towards the past is so strewn with pitfalls. Controversies are common and occasionally fraught, particularly when the aim of reconstruction is reintroduction. Proposals to bring back the white-tailed eagle to Wales, for instance, have delighted and alarmed in equal measure; for every nature-lover that would rejoice at the sight of these birds soaring beside the mountain crags, there is a farmer that fears for the fate of his lambs. In 2021, the Welsh Ornithological

Society printed a rebuttal to the claim that eagles were native, contesting that the reliance on place-names was too simplistic. Take the example of *Craig yr Eryr*, Rock of the Eagle, wrote Dewi Lewis: 'Does this mean that it is a rock where eagles were common or a rock where one eagle was seen and that this was considered an exception, a one-off event that needed to be recorded and remembered? Or a rock where someone believed they had seen an eagle? Without any firm supporting evidence as to the origin it is difficult to offer anything of substance.'[16] Researchers have also challenged the suggestion that beavers were native to the whole of California. Such claims, they wrote, were due 'in large part to the species' charismatic nature rather than by the presentation of sound evidence'. Furthermore, they argued that efforts to bring back the rodent were damaging California's genuinely native species.[17]

I don't know the truth in either case. Nobody does. Given the uncertainty of the evidence and the potential ramifications of reintroductions, it is worth taking both sides seriously. Even so, what is clear from these archives is the general trend: the natural world once contained a variety and abundance that is scarcely imaginable today.

We conceive of loss in binary terms – of the creatures that were there until they weren't. The Red List of Threatened Species, published by the International Union for Conservation of Nature (IUCN), categorises animals, plants and fungi according to how close they are to extinction. Given that nearly a third of species assessed are threatened with this fate, the focus from conservationists is necessarily upon saving the last of the most vulnerable. In instances where a species is still relatively abundant, conservationists are mostly happy to stay back: there are always more urgent tasks at hand. We ask not whether a species is thriving, but only if it is dying.

But this approach has put something else at risk of extinction: the language of plenty. These days, no one is likely to mistake a flock of birds for a tornado, as the ornithologist Alexander Wilson did when the sky above him darkened with passenger pigeons in the early 1800s. The species is now extinct.[18] No more do the children of London need to worry that a red kite might pluck the bread from their hands, as often happened during the 1500s, when the bird was so common it was considered vermin. Having been pushed to extinction once, the reintroduced birds now keep a wary distance, and sightings in inner London are rare.[19] Nor is any feast likely to include 204 cranes, as the banquet for the Archbishop of York did in 1465. Today, there are only around 200 cranes left in the whole of Britain.

We have become accustomed to thinking small – to accepting the threadbare scene before us, forgetting the splendour of the original weave. And so we celebrate the return of a handful of white-tailed eagles to the English countryside, forgetting that thousands of these birds would have been present in the past. The IUCN lists the beaver as a species of Least Concern, as populations are considered to be stable, ignoring the fact that numbers in North America have dropped from up to 400 million pre-settlement to around 10 or 15 million today.

This process of forgetting is so ubiquitous that scientists have given it a name: shifting baseline syndrome. The phrase was coined by fisheries biologist Daniel Pauly in 1995. It describes how the goalposts of nature are constantly moving: we assume that the landscape of our childhood was rich and wild, measuring any losses against that memory.[20] The next generation does the same – except this time, the landscape of childhood is already a little less rich, a little less wild. Although a human lifetime is generally long enough to notice the whittling away of the natural world, it is not sufficient to fully comprehend the kind of catastrophic crashes that become apparent when we delve back further in time. Eventually, these shifted baselines are transformed into targets, locking in the loss for the future. We aspire to depletion because we are too short-sighted to know what abundance looked like in the first place.

Happily, conservationists are beginning to overcome their amnesia. In 2012, the IUCN endorsed a new approach to conservation: as well as the Red List, there would now also be a 'Green Status of Species', measuring recovery against an historical baseline as well as extinction risk. For a species to be considered fully recovered, it must have returned across its indigenous range, to the point where it can perform its natural functions in the ecosystem. Such calculations can make the wildest dreams of conservationists seem like little more than tinkering. I asked Molly Grace, an ecologist at the University of Oxford who spearheaded the design of the Green Status, for an example of what such a recovery might look like. 'Maybe you reintroduce wolves to England, for example, but maybe there are only ten of them,' she said. 'Would we say that's a functional population of wolves? No.'

Ecologists spent almost a decade grappling with the complexities of creating such a metric, the most difficult being the baseline against which recovery should be assessed. They finally decided on a range of dates – between 1500 and 1950 CE –

recognising that humans have unleashed their destructive potential at different times across different parts of the globe.

Even this benchmark will only go so far towards redressing the loss of biodiversity that humans have caused. Somewhere like England, for instance, the metric is unlikely to prove very useful, given that wolves, beavers and bears had already vanished by the sixteenth century. Indeed, Eric W. Sanderson, a landscape ecologist at the Wildlife Conservation Society, argued that the Green List was only further embedding memory loss. 'Why should conservationists forget past extirpations because they happened an arbitrarily long time ago?' he wrote in the journal *Conservation Biology*. 'Lions, for example, were extirpated from the Greek Islands in ancient times, which may seem like a long time to human beings, but from the perspective of the 3.5-million-year evolutionary lineage of the lion, a 2000-year range loss is but a brief interregnum. If such a former reality seems divorced from contemporary circumstances, it is because baselines have shifted, not because lions could not live in Greece again if Greeks made space for them.'[21]

Imperfect though it may be, the IUCN Green Status is a good start to a task that, until recently, seemed impossible. Already, almost forty animals have been formally assessed. The findings illustrate the chasm that has emerged between the abundance of the past and what is considered healthy today. According to the Red List, for example, the American bison is currently doing well. According to the Green Status, however, it is 'critically depleted', with overhunting and habitat loss having drastically reduced populations since 1750 CE, when vast herds roamed the majority of the United States, Canada and Mexico.[22]

The Green Status is not meant to provide targets for conservation or policymaking. To recover species in large numbers across their historical range would be impossible without making major sacrifices. Certain species have declined because of luxuries we could have easily forgone – because we wanted to wear fur coats and dine on shark fin soup – or because their wildness rubbed up against our sense of civilisation. Others have been pushed to the margins for more reasonable requirements. The fields, cities and solar panels required to sustain an expanding population needed to go somewhere, and that somewhere was wilderness – it could not have been otherwise. Wildlife cannot be restored to prehistoric levels without massive reductions in the human population: the earth will only support so much life. The subjugation of the wild is the trade-off we make for the comforts that we enjoy today.

At some point, however, we began to take too much. Whether that point arose at the end of the Pleistocene with the extinction of the megafauna, or with the acceleration of impacts in the twentieth century, or somewhere in between, is always going to be a subjective call. But one thing is clear: measuring conservation against a baseline of extinction has allowed us to skirt the uncomfortable truth of this bargain for too long. We are not coexisting with nature, but rather obliterating it. Only by shifting towards an historical perspective may we realise the extent of the damage: instead of counting our successes, we are forced to confront the depth of our debt.

Not only have we forgotten the old abundance of the natural world; we have also lost sight of how wildlife behaved before their habitats were decimated by humans. The preferences we witness in the wild today are not those of animals with the luxury of choice: they are those of survivors that have been backed into a corner, making do with what little remains.

We tend to think of species as having a fixed abode, their guts tailored to a particular diet and bodies honed to a certain terrain. The working assumption among ecologists has been that, to reconstruct an animal's historical range, all you have to do is figure out the past extent of the habitat in which it lives today: more swamps would have meant more swamp dwellers, more woodland meant more woodland specialists, and so on. It has only been in the last decade or so that this belief has started to unravel. It was in Poland, at the margins of Europe's wildest forest, that the questions first began.

Bison have long been considered the arch-herbivores of the forest. There are no written records of the species ever having lived anywhere else. Hulking and hairy, so large that its bones are sometimes mistaken for aurochs', the bison looks almost incongruously primeval, as though it ought to have been wiped out with the rest of the megafauna. Yet, somehow, it escaped the onslaught, and continued to haunt the forests of Eastern Europe long after its Ice Age companions had been exterminated. In his *Poem about the Size, the Ferocity, and the Hunting of the Bison*, published in 1523, Nicolaus Hussovianus depicts the animal in its supposedly natural habitat: young calves leaping over fallen pines, lustful males trampling the ground and trembling the oaks, rotten wood threatening to trip the horses of those who dared to hunt them.

For, despite their size, hunted they were. Bison were the favoured quarry of royal hunting parties. Beaters would drive the animals into netted enclosures, whereupon the noblemen would take aim with long-barrelled harquebuses – the predecessor of the musket – after which servants would haul away the corpse by sleigh.[23] Accounts tell of medieval banquets where sixty bison were served at a time for days on end.[24] Such excesses, alongside exploitation of the woodlands for timber and potash, took their toll: one by one, populations began to fall from the map. By 1717, bison had vanished from Moldova. In 1755, they went extinct in Eastern Prussia. The last herd in Transylvania disappeared in 1790.[25] Soon, there were just two wild populations remaining, the most famous of which was in the Białowieża Primeval Forest in Poland.

The survival of bison in Białowieża did not happen by chance. In the fifteenth and sixteenth centuries, peasants were entitled to take hay from the meadows to feed their own livestock. The remaining stubble, and any leftover haystacks, provided unofficial fodder to the bison, keeping them fat and healthy when times were lean. Populations expanded in response to the ready supply of food. By the early 1700s, supplementary feeding had become official policy, endorsed by the monarchy, which was anxious not to lose access to Europe's biggest game. Russian tsars, who took on the protection of Białowieża after Poland lost its independence in 1795, intensified the practice. Inside the forest were barns and feeding racks, which servicemen would replenish with hay. If the men failed to provide, as sometimes happened in times of political turmoil, the bison would take matters into their own hands, feasting on the villagers' own haystacks 'mostly by assault', as one forester complained in 1797. This giant herbivore, a consummate survivor, was apparently incapable of making it through the winter without a helping hand from humans. By the nineteenth century, Europe's wild bison were not so far from livestock in their dietary requirements.

Ultimately, however, even this policy wasn't enough to keep them alive. After centuries of protection, World War I meant that the bison of the Białowieża were finally left to fend for themselves. German troops killed around 600 of the remaining animals and, by 1919, the population there also died out.[26] Europe's last wild bison was shot in 1927 in the Caucasus Mountains. Poaching and neglect had finally conquered the megaherbivore that the hunters of the Pleistocene had somehow left behind.

Before long, however, the bison were back. A handful of the creatures still survived in zoos across Europe. At the first International Congress of Nature Protection in 1923, a Polish ornithologist called Jan Sztolcman proposed their rehabilitation. Soon after, the International Society for Protection of European Bison was born; plans for breeding and reintroduction into the wild were top of the agenda. Unsurprisingly, given everything that naturalists knew of the species at that time, forests were seen as the prime habitat for their return.[27]

After decades of preparation, two captive-raised bison were released back into Białowieża in 1952. Over the following decades, more animals were returned to their old ranges across

Lithuania, Belarus, Ukraine, Russia and Kyrgyzstan.[28] In each case, it was a forest that welcomed them home. Even then, caretakers continued to provide the bison with hay and silage, supplementing their woodland diet of acorns, browse and bark, sometimes replacing it almost entirely during the harsh winter months.

In many ways, the reintroduction programme was a phenomenal success. The bison bred, calves were born, and numbers began to shoot up – first into the hundreds, then into the thousands. Europe's forests had their figurehead once more. But then, twelve years after they were put in the forest, the bison of the Białowieża decided they wanted out again. The golden meadows beyond the periphery of the forest were apparently more appealing than the primeval gloom to which they had so long been confined. The migration was slow at first, but then the number of nomads began to increase.[29] This was the point at which the reintroduction programme started to become too successful, at least in the eyes of the farmers, whose crops suddenly comprised a major part of the bison's diet. Compensation for the lost yield began costing the Polish government tens of thousands of euros every year, rising to more than €90,000 by 2010.[30] Nor were the Białowieża bison alone in their appetite for wandering. Almost 70 per cent of the populations that had originally been reintroduced to forests subsequently expanded into the farmland, meadows and marshland beyond.

It wasn't that Białowieża had reached saturation point: some areas remain unoccupied or little used. Nor was it for want of food, with feeding stations set up throughout the forest, just as they had been in the past. Nevertheless, some mysterious and irresistible force was indisputably drawing these forest creatures out from the darkness and into the light.

Seeking to better understand this peculiar behaviour, three academics started to delve into the deep past. It so happened that Graham Kerley was visiting the Mammal Research Institute, based at Białowieża. As a South African ecologist, used to seeing bovines – buffalo and so on – in the savannah, he wanted to understand 'why these European people have cows in the forest', as his collaborator, Joris Cromsigt, put it to me years later. Cromsigt, an ecologist from the Netherlands, was also working at the Institute at the time. Together with Rafał Kowalczyk, the Institute's resident expert in large mammals and forest ecology, the scientists published a paper setting out an alternative history of the bison.

To really understand the present-day behaviour of the bison, they argued, you have to look back to the Pleistocene. European bison, *Bison bonasus*, first emerged some time after the Last Glacial Maximum, around 15,000 years ago,[31] around the same time that the ancestral steppe bison, *Bison priscus*, went extinct. The steppe bison was a grazer: a creature of tundra and grasslands. The exact relationship between the two species – whether they existed in parallel or if one emerged from the other – is unclear.[32] But there seemed no reason to believe that European bison did not occupy the same open habitats as its predecessor. Its physical characteristics, including its wide muzzle and high-crowned teeth, all suggested a species adapted to eating grass rather than woody vegetation.[33] The bison, they hypothesised, was a refugee species: the ancient forests where they had survived for so long were not the final relics of their original habitat, but rather the last hideaway of a species on the run.[34]

In 2012, they published their findings. Other scientists were quick to expand upon their claims. One group was able to reconstruct, with startling precision, how the bison's diet had changed across the millennia. By analysing bone samples, they found that, in the early years of the Holocene, bison fed in mainly open habitats. This made sense: the glaciers had only just retreated, leaving open tundra and shrubland in their wake. Alongside grasses and shrubs, the bison's diet also included leaves and lichen, indicating some forest feeding too, showing a degree of flexibility that would soon prove useful as the wildwood began to spread across the continent. As grasslands became swamped by wood, the bison retreated into the darkness, switching its diet accordingly. But the wildwood, as we have already seen, was on borrowed time. Not long after the bison had made their move into the forest, the first farmers began to open up the landscape once again, providing the perfect opportunity for the species to reclaim its preferred habitat.[35]

Except it didn't. Coexistence with humans was no longer possible. Fields of crops were incompatible with the heavy hooves and big appetites of this herbivore, and it was obvious which species had the upper hand in any conflict. And so the bison stayed in the forest, making do with increasingly fragmented patches, hanging on in low densities, and later in higher densities with the arrival of medieval hay provisions. It was only in the last seventy years, as their status changed from royal quarry to conservation priority, that the bison felt ready to risk re-emergence, sloping out of the forest and across the cultivated remains of their ancestral steppe.

Twentieth-century conservation may have saved the bison from extinction, but by restricting the species to the forest, its saviours had once again fallen victim to shifting baseline syndrome. This time, the collective amnesia concerned not abundance but rather habitat and behaviour. No one could remember what a truly wild bison was really like.

Already, the years in exile have taken their toll. Inbreeding and low genetic diversity were always going to be a problem among the reintroduced bison, given that the population stemmed from a handful of captive individuals. But these problems have been exacerbated by their confinement in a suboptimal habitat as, without supplementary feeding, populations will always remain artificially low. Meanwhile, the congregation of animals around the feeding stations in winter has led to the spread of a bloodsucking nematode, leading to chronic diarrhoea, deterioration and even death among the calves.[36]

Crucially, the failure to consider open areas for bison reintroduction limits the area of land where this threatened creature can make a comeback, particularly in the west, where woodland is scarce. Indeed, a myopic focus on forest means that the potential of abandoned farmland as a future habitat has been overlooked. As we saw in chapter four, projected trends mean that this will cover a larger proportion of Europe's marginal land in the coming decades. The reintroduction of bison could prevent these former fields becoming overrun by trees, instead creating a mosaic of grassland, groves and scrub, and thus preserving the many wildflowers and insects that thrived among the crops. Meanwhile, the return of a charismatic mammal could catalyse opportunities for tourism, providing an alternative source of income for local communities.

Despite the wealth of historic evidence, however, the refugee species hypothesis has been difficult to test. The health of forest-based bison cannot be compared to grassland-based bison because the latter does not exist. For decades, the assumption that bison was a forest species was so ingrained that no one had even considered putting it into an open landscape – with one exception. In April 2007, a handful of bison were reintroduced to a small dunescape, called Kraansvlak, in the Netherlands. They have lived there ever since.

Kraansvlak belongs to the Dutch water company, PWN, which was looking for ways to heal the damaged biodiversity of the dunes. In recent decades, several species of birds, including stone curlew and red-backed shrike, had disappeared entirely, while others had suffered steep declines, following years of acidification and over-fertilisation. The problem was exacerbated by the decimation of the local rabbit population – for centuries the primary grazer on the dunes – by an outbreak of myxomatosis in the 1990s. Their nibbling had helped to keep the dunes open and sandy despite the deposition of excess nutrients; when they disappeared, coarse tall grasses expanded unchecked, drowning out the rarer and more delicate species, including lichens and mosses.[37] The once rich and shifting landscape had become matted and stagnant.

Highland cows and Konik ponies had already been brought into the area in an attempt to break up the vegetation but to no avail. Bison were the natural next step.[38] Poland had recently joined the European Union, meaning the prospect of sourcing and transporting the creatures into the Netherlands posed slightly less of a bureaucratic headache. Still, the idea of returning bison to a small, fenced reserve, in the densely populated Netherlands, was a radical one. The image of this snorting primeval beast mooching around on a beach near Amsterdam seemed patently absurd. Moreover, the team decided that they wanted to experiment with allowing the bison to fend for themselves. That meant no supplementary feeding: if the bison wanted food, they would have to find it amid the dunes. It would be a rare opportunity to see how the animals behaved when left to their own devices.

Although the refugee species hypothesis had yet to be published in the academic literature when the bison were brought to Kraansvlak, the idea was already floating around among practical-minded ecologists. At a workshop in 2004, bringing together experts to discuss the proposal, the ecologist Han Olff suggested that the flora of the dunescape was designed to deal with absent oversized creatures. 'Why do sea buckthorns, hawthorns and roses have spines? Why does elderberry leaf taste so bad? There are many characteristics of plants and animals that can clearly be traced back to adaptations to large herbivores.'[39]

Three years later, the first three bison arrived from Poland. They were transported in wooden boxes, carried into the reserve on the back of blue trucks. Upon their release, they headed straight to the top of the first dune they saw, before disappearing into the shrubbery. Henceforth, the only indication that these were not fully wild creatures were the GPS collars hung around their necks, providing hourly updates on their locations. Just under a year later, another three bison were brought in to join the herd.

Once settled in Kraansvlak, the bison had the pick of a rich selection of habitats. Alongside the open sand of the dunes were expanses of shrubbery, forest, grassland and wetland. Cromsigt, one of the three scientists behind the refugee species hypothesis, was brought on board to lead the ecological modelling. For the next ten years, he and a team of researchers kept tabs on the animals from afar, watching what they ate and where they chose to live.

At first, the bison seemed timid in their new surroundings, gravitating towards the deciduous forest. Within three months, however, they had emerged, spending a large part of the summer between the dunes, with grasses ultimately making up around 80 per cent of their diet. Not even the winter snow hampered their appetite for grazing: like the plains bison of North America, they would simply sweep it away with their muzzles and hooves, making craters to gain access to the nourishment beneath. At times the bison added bark and acorns foraged from the woodland to their diet, but at no point was there any need to bring in any food from the outside world: the enclosure provided everything they needed for year-round survival.

Has the experiment proved the hypothesis that the bison is a refugee species? I put the question to Cromsigt. He demurred slightly: the Kraansvlak experiment was small, and there have been few efforts to replicate it. But, overall, the outcome has been positive. 'You can never prove a hypothesis in that sense, but we found a lot of evidence to support it,' he said. 'Their diet is actually quite similar to species like horses and cattle, suggesting that they are not a typical forest species, because then you would expect a bigger diet difference. What we have seen is they have been very reproductive. They still produce quite a few calves every year. If you take that as a measure of fitness, it suggests that they are doing very well in this paradigm.'

Almost two decades after the bison first wandered across the dunes, the standard approach to reintroduction still involves supplementary feeding, making it difficult to assert anything more definite than that. Even so, the evidence base is growing. Over the past decade, more herds have been fitted with GPS devices, allowing researchers to build real-time maps of their movements. Scientists have found that each of Ukraine's seven herds make use of open environments, including meadows and crop fields. The Uladivska herd, for instance, spends the early summer days deep in the hornbeam forests, avoiding blood-sucking insects, drifting towards open glades each morning and evening to forage; in late summer, when the insects are less voracious, it ranges further outwards, feasting on fields of oat, corn and beet. In the Carpathians, meanwhile, the Bukovinska herd gravitates towards the meadows of the foothills in the spring, before moving up the mountains to feed on the wild roses, bilberries and strawberries of the higher elevation hayfields during the summer. All of these herds received supplementary feeding during the winter, and it was during these months alone that they remained sedentary within the forest.[40] Studies of free-living populations in the Czech Republic and Lithuania have painted a similar picture: that bison incline towards open land, but live happily enough among the trees when they have cause to do so.[41]

Shifting baseline syndrome and the refugee species hypothesis are surely the two most depressing concepts in ecology. They are a reminder that even our moments of optimism, our success stories, our most ambitious targets, have failure baked into their hearts. It cannot be otherwise: so long as humans remain dominant, nature can never return to its primeval abundance. And yet, instead of allowing that baseline to inch forward, we can learn to push it back. Perhaps, if we move over just slightly, we could rewind to a more recent past, to a time when animals were afforded a little more space: to the baseline of our grandparents, of mountain men, of medieval feasts.

SEVEN

The Laboratory of Time

Now we travel back in time by some 56 million years. We are in the epoch known as the Palaeocene. It has been 10 million years since the dinosaurs were wiped out by an asteroid, and Planet Earth looks both strangely familiar and totally unrecognisable.

Life rebounded quickly following the collision. Trees, flowers and animals re-emerged in uncanny doppelgangers of the ecosystems that we know today. There were redwoods in the Arctic and forests in Antarctica. Incomprehensibly giant snakes slithered through the tropics of South America. Flightless birds, so large it was once thought they ate horses, loped through France. Mammals, no longer living in fear and competition with the dinosaurs, diversified into myriad forms, none of which have survived into the present.[1] After so many years of dust and darkness, the world was green, luscious and vibrant once more.

Then, and no one quite knows why, a huge plume of carbon dioxide entered the atmosphere. The episode that followed was called the Palaeocene-Eocene Thermal Maximum, or PETM, and it caused global temperatures to rise abruptly by between

5–8°C – to an average of around 23°C compared to 15°C today. For some 200,000 years, the earth was a hothouse.

The heat sent the post-dinosaur world into turmoil – again. The single-celled deep-sea organisms, known as benthic foraminifera, suffered the most: they went extinct en masse. Other creatures moved and adapted to deal with the change. Mammals grew smaller. Primates and hoofed mammals spread into North America across the land bridge that linked it to Eurasia. Near the North Pole, the ocean warmed to around the temperature of a swimming pool and flowers bloomed across the land.[2]

Let's zoom in now to Wyoming, where the next instalment of this story takes place. During the Palaeocene, the state was warm, wet and swampy. Turtles and crocodiles lurked among forests of bald cypress, witch-hazel, walnut and birch. Although there is no exact analogue in the present day, the landscapes of the Carolinas come somewhere close. Following the spike in temperature, however, this ecosystem changed abruptly. The once-humid climate became dry and tropical. The existing tree species vanished, replaced by an assemblage of more southerly flora spreading polewards: an explosion of ferns, palms and beans, including poinsettia, sumac and paw-paw.[3]

We know all this because the fossils of these plants have been preserved to this day. Most leaves, when they fell to the ground, rotted back into the soil. But some found themselves on a different journey. After wafting through the warm air, they instead came to rest on the calm surfaces of the many ponds and other water bodies that dotted the basin. Here they floated momentarily, before sinking to the bottom and becoming buried by mud.

Over time, their shapes became imprinted on the rocks, preserved forever in a palette of rust, in patterns as delicate as any doily.

And that's where they remained, undisturbed for millions of years, until a scientist called Ellen Currano arrived with her chisel and pick-axe to bring them to the surface once more.

'Palaeobotany is quite different from dinosaur palaeontology, which is what most people have seen on television,' says Currano, now an associate professor at the University of Wyoming. Instead of scouring the rocks for that one perfect skeleton, palaeobotanists will typically uncover thousands of specimens at a single site. There may be fewer eureka moments, but what is revealed in the rocks is marvellous in its own way. Rather than uncovering a single animal at its point of death, what hunters of plant fossils reveal is a snapshot of an ecosystem at a particular moment in time.

The impacts of the PETM are particularly well preserved in the rocks of the Bighorn Basin in Wyoming. The fossils are close to the surface, rendering the place a palaeobotanist's paradise. Currano had noticed that, as temperatures spiked, the appearance of the fossils underwent a sudden transformation. It was not only the plant species that changed, but the condition of the leaves themselves. They were less intact, less perfect, more holey. In these rocks was written the 56-million-year-old prequel to *The Very Hungry Caterpillar*.

There are many ways in which an insect can damage a leaf. Some bite chunks from the centre, while others nibble around the edges. Some pierce and some suck. Some will eat everything but the veins, leaving behind fragile skeletons, or cause galls to form where they lay their larvae. Generally speaking, the greater the variety of the damage, the greater the diversity of the species causing it. And, during the PETM, the diversity of the damage increased unmistakably. The fossils of this period bear a variety of battle scars. Some appear to have lost a fight with a hole-puncher while others look as though they have been hit by a cannonball. In some cases, insects have left inky trails across the rocks, creating the impression of old treasure maps. Some look as though a cigarette has been stubbed out on their surface.

For Currano, the implications were clear: the warmer temperatures of the PETM had caused an influx of new insects into the Bighorn Basin. The tropical plants had been accompanied on their journey by a pageant of tropical invertebrates.[4]

The damage was not only more diverse; it was also more frequent. Almost 60 per cent of leaves fossilised during the PETM bear the marks of an insect, compared to 15 to 38 per cent during the cooler years of the epoch that preceded it. It is possible there were just more insects around at this time. Or perhaps the cold-blooded invertebrates became hungrier as the climate grew hotter. Maybe the leaves contained fewer nutrients as the chemistry of the atmosphere changed. Whatever the reason, those appetites left their mark. Climate change caused the trees of this period to disappear at an unprecedented pace into the bodies of its insects.

Fast forward to the present day, and the climate is changing once again. This time, the cause is no mystery. Deforestation, agriculture and the combustion of fossil fuels are heating the planet at an alarming rate. Temperatures are almost 1.5°C higher than they were in pre-industrial times. The PETM was a natural experiment into what happens when you release vast quantities of carbon dioxide into the atmosphere. Today, we are conducting an unnatural one.

Currano likes to refer to the PETM as humanity's 'best case scenario' – a window into our strange and sweltering future. The thousands of petagrams of carbon that were responsible for the PETM temperature spike were released over thousands of years. The best estimates suggest that modern society is now emitting carbon at around ten times that rate. Upon this basis, the palaeontologist Philip D. Gingerich published a paper in 2019 in which he calculated that the accumulation of carbon would reach PETM levels 'in as few as 140 years or about five human generations' – with the climatic upheaval to match.[5]

The ecological revolutions of the deep past were all part of the chaotic story of our planet: the random events that led to the evolution of the world as we know it today. The same is not true of the current episode of disruption: our species depends upon the stability of the natural world for its survival.

The migration of insects into Wyoming is a case in point. During the PETM, the invertebrates arrived alongside the vegetation that supported them. Today, a similar invasion could cause a food security crisis. The route from the equator is littered with human obstacles. While beetles, butterflies and ladybirds might be able to fly over the highways, cities, megafarms and strip malls, the plants would find it much harder to migrate. The animals that once dispersed their seeds are gone, as are the soils where they would once have put down their roots.

That means that the insects will be forced to turn elsewhere to satisfy their appetites. To the crops intended to feed America's human population, perhaps. Currano's research suggests that, if the pattern of increased consumption remains the same as it was 56 million years ago, then farmers can expect big losses in the future.

'One of the very immediate lessons is that we can expect more munching to occur,' she says. 'So how are you going to offset those losses? Are you planting more? Do we need to be devoting more land for agriculture? I think a lot about how plant diversity affects how much feeding is occurring, and if you have these giant monocultures, are you going to see even greater herbivory losses than you would in a more mixed environment? I don't know what the answer is.'

Meanwhile, existing ecosystems may struggle to cope with the unfamiliar bugs, against which they have evolved no defences. 'You potentially have plants being exposed to insects that they have not previously been exposed to,' Currano adds. 'Maybe they are defended against these particular insects and that doesn't matter. But maybe they are not.'

This book has explored how, by looking to the past, we can restore something of the earth's lost abundance to the depleted landscapes of the modern day. So far, I have restricted this discussion to recent history – by which I mean the past 100,000 years or so.

In taking this perspective, I have echoed most of the published literature on the topic, as well as the interests of conservationists. Models for restoration rarely delve back any further than the Holocene. Those invested in the megafauna sometimes turn for inspiration to the most recent interglacial of the Pleistocene. When it comes to anything before that, however, the science is scarce. That is because the Holocene and late Pleistocene are considered the 'natural' baselines of our modern landscapes – and for good reason. By these points, geologically and ecologically, the planet had more or less reached an equilibrium that, without human interference, would have continued to the present day. Any subtraction from these baselines is considered a diminishment, while those who propose additions are accused of reducing 'real' nature to little more than a zoo. Climatically, however, we are now headed down a different path. To truly understand the destination, we must peer even further into the past.

It is easy to see why we have been so short-sighted. 'The past is a foreign country: they do things differently there,' wrote the novelist L.P. Hartley in 1953. The further back you go, the more acute those differences become. Time itself becomes difficult to comprehend. When I contemplate the billions of years that have passed since the earth began, my mind begins to warp in a similar fashion as when I imagine the miles between me and the edge of the ever-expanding universe. The -illions start to blur and numbers lose their meaning. Picturing the ecosystems of those years, meanwhile, demands an enormous stretch of the imagination: it requires us to conjure plants and animals with no equivalent today; landscapes in which plants and animals did not even exist. To erase the oceans, rip down mountains, tear continents asunder. To grow forests where it seems forests would never grow; to locate animals where it seems they would never live. The deep past undermines any sense we might have that the world is knowable or stable. We live within a pause in the anarchy, upon a planet where nothing truly belongs anywhere.

And so it is easy to write off those eras as irrelevant to the modern world. How does it help the polar bears to know that redwoods once grew in the Arctic?

But another twist of time's kaleidoscope is now underway – and, this time, it is humanity that is rotating the dial. Climate change is causing the next round of pandemonium. Even if we had the option of restoring the ecosystems of the early Holocene or the Last Interglacial in all their finery, there is no guarantee they would survive the tumult that lies ahead. Already, at less than 2°C of warming, coral reefs are bleaching, forests burning, glaciers melting, species moving.

People refer to the current climate crisis as unprecedented, and, in some ways, it is. It is the first time that a single species has been solely responsible for upending the stability of their own epoch. In another sense, however, the changes that we are witnessing today are nothing unusual. The PETM is just one of many periods when the earth was hotter than it is today. Mind-bending as it may be, the deep past can act as a natural laboratory for understanding the world that lies ahead as we hurtle further into the Anthropocene.

The idea of looking into the deep past to predict the future has been around almost as long as the concept of climate change itself. NASA scientist James Hansen threw the greenhouse effect into the global spotlight for the first time when he testified before the U.S. Congress in 1988. Two years later, the UN's climate science panel published its first report assessing the evidence for global warming. One year after that, the palaeoclimatologist Thomas Crowley published a paper that examined the various warm intervals that have occurred over the last 100 million years to see whether any of them might be suitable analogues for the climate of the future.[6] Broadly not, he concluded: in none of the fluctuations that had occurred to date was there a moment that precisely replicated the conditions predicted for the twenty-first century. 'From a geologic perspective,' he wrote, 'mankind will be sailing into the future on uncharted waters.'

More than 30 years later, scientists remain more or less united on this point: the fossil record is not a crystal ball. Even so, Crowley wasn't arguing that the deep past was totally useless when it came to predicting the future, but rather warning against taking it too literally. Meanwhile, in the three decades since he published that paper, palaeoecologists have become better at peering through the fog. There have now been several attempts to piece together the information within the strata into something resembling a guidebook to the years ahead.

The PETM is one of the best analogues for the future because the rise in temperature, like today, was caused by the emission of greenhouse gases. Moreover, the volume of carbon released was roughly equal to that contained within existing fossil fuel reserves. However, the comparatively slow rate at which this carbon was released into the atmosphere gave species time to adapt and relocate. This meant that, as one study put it, the ecological disruption that occurred should be seen as 'a lower estimate of the potential future impacts'.[7] In addition, the flora at this time was unlike that of today: the ancestors of those ancient plants had evolved in the hothouse of the late Cretaceous period, which may have primed them for the tropical weather to come.[8]

As such, some scientists have suggested that the more recent past, when ecosystems looked much like they do now, makes for a better model of the future. The transition from the last Ice Age of the Pleistocene into the warmer landscapes of the Holocene has been held up as an example of how climate change might play out through the twenty-first century and beyond. Temperatures rose quickly at this juncture, at around 0.17°C per decade on average, holding a mirror to the rate of 0.2°C per decade that is unfolding today – although that figure is accelerating fast.[9] Some have argued that this makes it particularly helpful for predicting how nature will respond to future heat. But there is one major objection to using that period as an analogue: the ecosystem at this time was no longer pristine. The loss of the megafauna at the hunters' hands, as we saw in chapter one, was already impacting everything from forest growth to wildfire frequency to seed dispersal. Unlike earlier epochs, the consequences of the climate cannot be easily disentangled from those of humanity.

And so it is to the deeper past that we must return to find a comparison. Obviously, no one knows exactly how the future will play out. Perhaps politicians will suddenly unite and agree to an immediate and dramatic reduction in emissions, or perhaps the opposite will happen and greenhouse gas emissions will start to accelerate once again: nothing is guaranteed. In 2018, a group of scientists published a paper that attempted to encompass these uncertainties.[10] They found that, in a worst-case scenario, the climate will resemble that of the Eocene – the epoch that followed the PETM – across much of the globe by the middle of the next century. This is a terrifying prospect. The Early Eocene was the warmest sustained period since the dinosaurs, when temperatures were around 13°C higher than they were at the end of the twentieth century: a time when the poles contained no permanent ice and rainforests grew in Antarctica.[11]

Should humans succeed in stabilising emissions before the end of the century, then the outcome looks less dire, although it is hardly cause for celebration. Rather than recreating the conditions of the Eocene, we will merely be entering into a world that resembles the mid-Pliocene, around 3 million years ago, when temperatures were up to 3.6°C higher than during the late twentieth century. During this epoch, sea levels rose up to 35 m above their current levels – an increase that would swamp many of the world's major cities and displace millions from coastal areas.[12]

Of course, you don't need to look millions of years into the past to know that climate change will shake up the natural world. The kind of insect migrations that affected Wyoming during the PETM are already happening today, to a lesser degree. Just re-visit the newspapers of 2019, when the discovery of a so-called 'murder hornet' in Washington State led to a spate of headlines across the nation. The wasp, which can measure up to 2 inches long, is a threat to honeybees; it can wipe out an entire hive in hours. Experts believe that warmer temperatures have allowed them to expand from their native range in east and southeast Asia into the United States.[13] In Europe, oak processionary moths are expanding northwards, where the once inhospitable climates have started to become increasingly tolerable, with regular outbreaks now taking place across the south of England.[14] These insects strip oak trees of their foliage, leaving them vulnerable to other stressors, and can cause rashes and breathing difficulties for humans.

However, while the principle that species will shift and expand in response to climate change is well established – indeed, observable in real time – forecasting these movements is surprisingly challenging. To do so, scientists depend upon a combination of recent observations, controlled laboratory experiments and statistical models. But these methods provide no more of a crystal ball than do fossil records: they are plagued by bias, or focus on one species alone, or fail to consider nature's tendency to respond to change in messy and unpredictable ways. Dependent on records from the past few decades, these models become increasingly untested the further they attempt to project into the future: observations from the last fifty years are a poor indicator of what plants and animals might do centuries, or even millennia, from now.[15]

No one is claiming that these models are useless, but rather that they can only ever tell us half the story. By looking into the deep past – murky as the view might be – scientists can examine how ecosystems actually responded to periods of upheaval, similar to that which is coming our way. The fossil record makes it possible to identify consistent patterns across huge swathes of time and space that cannot be predicted from statistics and recent observations alone: an opportunity to find precedent amid unprecedented times.

Talk of insect invasions and Antarctic rainforests may sound alarming. But amid the doomsday scenarios is a message of hope. Count the extinction events caused by global warming alone in the deep past and the answer is: very few, if any. The overriding pattern of the past 66 million years has not been collapse but change.

Let's return to the PETM. Some 200,000 years after the release of carbon that turned Wyoming's swamps into tropical savannah, global temperatures plummeted back to previous levels. The natural world reassembled once more – and the ecosystem that emerged in the Bighorn Basin turned out to be remarkably similar to that of the previous epoch. As rain re-wetted the floodplains, back came the walnuts, the birches and the cypresses. Mixed among these trees were some of the immigrant species that had arrived in North America during the episode of extreme heat. The landscape was not identical to that which had come before, exactly, but neither was it diminished. The PETM had caused no major extinctions or loss of diversity.[16] And the Bighorn Basin was not unique in this regard. Across the world, the majority of species – excluding benthic foraminifera – found a way to ride out the heat.

During the balmy years of the Eocene, it was a similar story. Contrary to suggestions that future climate change could turn much of the world's tropical rainforests into savannah, this was a time when that biome was significantly more extensive and diverse than it is today.[17] The same resilience was a feature of the Pleistocene. Despite the repeated and dramatic swings in temperature during this epoch, only one species of plant is known to have gone extinct: the Critchfield's spruce.

None of which is to say that no plant or animal ever disappeared in the deep past. Clearly, they did. It would be remiss not to mention the five mass extinction events that have taken place during the last 450 million years, some of which were associated with rapid warming. These, however, were periods of general chaos and upheaval, including the toxic impacts of volcanic eruptions: climate change cannot shoulder all of the blame. Notably, the most catastrophic of these events saw a similar magnitude of temperature rise as the PETM, when, as we have seen, extinctions were rare.[18] Many species, meanwhile, have been erased for reasons unrelated to the climate: because they evolved into something else, for instance, or because of an unfortunate genetic mutation. Where climate change has been implicated in extinctions, it is more often the result of cooling than warming. For millions of years, the natural world has shown itself to be remarkably resilient in the face of extreme heat.

This contradicts what we are often told about the climate crisis today. Many scientists have attempted to estimate how many species will go extinct due to future warming, and have come up with widely differing figures. For the most part, however, the prognosis has been devastating. One paper, published in 2015, concluded that around 8.5 per cent of species would disappear at around 3°C of warming, rising to 16 per cent beyond 4°C.[19] In another widely-cited study, published in 2004, the outlook was even worse: it estimated that up to 32 per cent of species would go extinct if temperatures rose by just 2°C, even if there was nothing to limit their capacity to disperse into new territory; that figure rose to 52 per cent if dispersal became impossible.[20] According to these predictions, we are not just looking at the loss of some benthic foraminifera; we are marching into a world that has been scrubbed of its life.

Scientists have come up with a name for this apparent mismatch between past trends and future predictions: the Quaternary Conundrum.[21] Based upon the models that generated these forecasts, the Pleistocene should have also witnessed widespread losses as glaciers advanced and retreated across the land. But it didn't – at least, not until *Homo sapiens* arrived on the scene. Why not?

Palaeoecologists have suggested that species have more coping mechanisms than we give them credit for. The vast majority of studies looking at future biodiversity loss are based upon models that hinge upon the territory currently or recently inhabited by the species in question. These conditions – in particular, the climate – are assumed to be the extent of what a species can tolerate. If those conditions cease to exist, so will the species. However, as anyone who has ever visited a botanical garden will know, this is not always the case. Behind the cast iron gates of Kew Gardens, for instance, you will find an arboretum filled with trees from across the world growing quite happily in the drizzle of south-west London.

What is true of these gardens is also true in the wild. Fossil records show that, in the past, many species occupied much larger territories, encompassing a wider range of climatic and ecological conditions, than they do today. Perhaps, somewhere on their journey of recolonisation following the last Ice Age, something stopped their spread: overharvesting, habitat destruction, or a mountain range, for instance. Or perhaps their journey is not yet over, and these species are still inching their way into the landscapes of the Holocene; still feeling their way forward through the warmth of the post-glacial world.

Either way, just because a species no longer inhabits a certain place in the present, there is no good reason why it shouldn't recolonise it in the future, either on its own steam or with a helping hand from humanity. One study into the fossil records of the jaguar across the Americas found that, during the Pleistocene, the species survived in a broader range of conditions that its descendants do today. Incorporate these old bones into the models and suddenly the impacts of climate change appear far less devastating: parts of Brazil that ecologists had believed would become unsuitable turn out to be acceptable after all. The ancestors of this big cat had seen it all before. In this case, the authors wrote, the more optimistic scenario was also the more realistic.[22]

If a species truly cannot tolerate the new conditions created by climate change – or is prevented from moving into new territories – there are still other ways to survive. It is worth remembering that every living thing on the planet today has an ancestor that endured the heat of the Eocene. Resilience is a feature, not an accident, of modern ecosystems. Some species are able to physically adapt to their new reality. This can happen over evolutionary timescales, through the process of natural selection, resulting in a shift in genetic makeup. As it evolved from its African ancestors, for instance, the woolly mammoth literally restructured its blood cells, enabling it to thrive in the freezing landscapes of the north.[23] In fact, recent evidence suggests that evolution is not the slow and chugging process it has long been thought to be, and that many birds and mammals can change surprisingly quickly – something that bodes well for their continued existence.[24] Other adaptations are more short-term, with animals adopting new habits and appearances to deal with their altered circumstances, much as I might gravitate towards shade or wear shorts to cope with hotter summers. Already, we are seeing signs of such flexibility: butterflies losing their spots to improve their camouflage within dry grass; tawny owls turning brown as snowy winters decline; cows producing less milk during warmer weather.[25]

For species unable to change or move in response to warming weather, there is another tool left in the box: retreat. Many species have survived episodes of climatic disruption by withdrawing to small pockets of safety, known as refugia, from which they could expand once conditions became favourable again. Broadly, during the frozen millennia of the ice ages, warm-adapted species endured by hiding out in the south, while cold-adapted species saw out the warmer interglacial periods by migrating towards the poles. However, it is likely that islands of climatic stability also occurred in more unexpected places, known as cryptic microrefugia, as a result of certain features of the landscape. Almost impossible to detect in the fossil record, these may have enabled the speedy recolonisation of certain species across the continents as the world began to tilt back to their preferred state. Following the discovery of bones in caves carved out from the hills in Devon, scientists suggested that such sheltered valleys may have offered refuge to red deer in Britain during the last Ice Age.[26] Indeed, it may have been the higher altitudes of the Rocky Mountains, on the boundary of the Bighorn Basin in Wyoming, that allowed the earlier vegetation to rebound so quickly at the close of the PETM, with the cooler, wetter climes offering a place of safety when temperatures spiked.[27]

Studies that consider these various adaptations tend to take a sunnier view of the future. A study of salamanders found that the number of populations destined for local extinction dropped by 72 per cent when their ability to adapt physically and behaviourally to heat was taken into account.[28] A study of forest bats, meanwhile, found that each population was adapted to the precise conditions in which it lived, rather than being monolithically suited to a certain climate. This raises the prospect of 'evolutionary rescue', where warm-adapted individuals spread their genes to those in cooler regions, dramatically increasing the species' chance of survival as the planet heats up.[29] A myopic focus on recent time means that we have overlooked the many strategies that present-day species would have deployed, through millennia of tumult, to make it into the twenty-first century. And that is good news.

Indeed, if the devastating predictions of loss are true, they have been slow to play out. To date, almost no creatures have gone extinct on a global level thanks to the current spike in temperatures. In Yosemite National Park, a group of scientists recently repeated a survey of small mammals undertaken by the biologist Joseph Grinnell between 1914 and 1920. Protected since 1890, Yosemite has been almost uniquely unaffected by the developments of the twentieth century, creating an unparalleled opportunity to tease apart the impacts of climate change from incursions such as roads, agriculture and pollution. Despite an increase in temperature of almost 4°C – far beyond the global average – the scientists found little change to the overall diversity within the park. None of the mammals surveyed had gone extinct over the past 100 years, not even locally. Instead, they had found ways to survive, with some – though not all – tracking cooler conditions by moving upwards into the mountains.[30]

It is rare to find reasons for optimism when writing about climate change. Such glimmers can be regarded with suspicion. Palaeoecologists have worried that, by highlighting the resilience of the natural world throughout the deep past, they will be accused of climate denialism or undermining the urgency of the situation. But the lesson from the past is not that everything will be fine. It is that species have somehow found a way to cling on amid the upheaval. To resist global extinction, even as they underwent widespread declines, local extirpations, evolutionary adaptation and dramatic shifts in range. By understanding these responses, we can better prepare ourselves for the challenges ahead.

'For me, the reason to study the palaeo-record is to help us to do the triage, the risk assessment,' says Jacquelyn Gill, a palaeoecologist at the University of Maine. 'A triage nurse in World War I would walk into a tent and scan the room and say: that guy's going to be fine, but this guy needs immediate attention. They would never say: people have survived battlefield wounds in the past and therefore we don't need to pay attention to anything in this tent.'

One man who has performed that triage is Mike Archer, professor of palaeontology and one-time director of the Australian Museum.

'I'm probably regarded as a bit of a maverick,' he muses, when I catch up with him over the phone from his home in Sydney. This is not an exaggeration. During our short conversation, he endorses a dizzying array of ideas for saving the world's endangered species, most of which would drain the blood from the face of the average ecologist. Introduce polar bears to Antarctica. ('Make them bipolar bears. Yes, they will gobble up a few penguins, but the population can handle that.') Let Richard Branson release Madagascan lemurs onto his private island in the Caribbean. ('There's one lizard that might be a bit annoyed about this.') Keep native species as pets. (Members of his personal menagerie have included a western quoll and a swamp wallaby.)

But what really animates Archer is the fate of the mountain pygmy possum, *Burramys parvus*. For this little rat-like creature, he has decided, is a soldier in need of immediate attention.

The cause, on the surface, is rather less glamorous than some of his other ideas – certainly less headline-grabbing. An unprepossessing marsupial whose only distinguishing feature is a saw-shaped premolar tooth, the mountain pygmy possum lives in the alpine zones of just three mountains in western Australia, where it survives the winters by hibernating under boulders. Counterintuitively, global warming is now freezing the species to death. During the winter, the snow insulates their rocky hideouts from the sub-zero air, heating them to a cosy 2°C. If the snow ceases to fall, however, the possums awake from their winter sleep and start to shiver to maintain their core body temperature, costing them vital supplies of energy. And that is what is happening as temperatures rise. The mountain pygmy possum is currently listed as critically endangered; a couple of bad winters and they could disappear forever.

The irony of their predicament is heightened by the fact that the mountain pygmy possum has been declared extinct once already. The first anyone knew of this little marsupial was when its fossil remains, probably dating from the Pleistocene, were discovered in a cave in New South Wales in 1894. At this point, it was assumed that the species had gone the same way as the woolly mammoths. But, in 1966, a live specimen was discovered at a ski lodge in Victoria: *Burramys* was back from the dead. This should be a story of redemption. Instead, it is quickly becoming a cautionary tale of squandered second chances.

Not if Archer has his way. For the past twenty years, he has championed a characteristically outlandish solution to the mountain pygmy possum problem – one that involves going 26 million years back in time. *Burramys parvus* is the latest in a long lineage of these little possums, stretching from the Oligocene to the present day. There have been four species in total, though never more than one at a time, suggesting that each one simply evolved into the next. The first three species of *Burramys* were essentially indistinguishable from one another. However, the latest – and possibly the last – stands out. It is the only incarnation to live in the alpine zone.

The ancestors of the present-day *Burramys* lived not on the mountaintops but in the temperate lowland rainforests. That the marsupials now find themselves stranded in this suboptimal snowy environment is, according to Archer, just an unfortunate accident of prehistory – one example of a creature that is failing to make full use of its potential range. During one of the warm intervals of the past, he argues, the wet forests of the lowlands were able to creep their way into higher altitudes, taking the pygmy possums with them. Here the possums made a life for themselves, hibernating beneath the rock piles and gorging upon bogong moths in the spring. When cooler temperatures forced the forest back down the mountains, the possums stayed put, making do with the sparse beneficence of the boulderfields. It was a fragile balance, but one that worked well enough for thousands, if not millions, of years – until climate change started tipping the scales. 'They are just hanging on in that area, and probably have through much of the late Pleistocene, by the skin of their little possum teeth,' says Archer.

His hypothesis is that the mountain pygmy possum has retained, to this day, an ancestral memory of its ancient forest habitat; that the knowledge necessary for survival in the lowlands remains embedded in its genes. He is confident that, with a little prompting, these instincts could be revived, allowing the *Burramys* to reinhabit the cool, lush landscape that served its predecessors so well.[31] However, with no corridor between the boulderfields and the forests that would enable it to make the journey itself, the possum will need a helping hand to make its way back down the mountain.

'Before we write this possum off, we should think about trying to get it back into the habitat that kept all its immediate ancestors very happy,' says Archer. 'At first, I was regarded as a bit of a nutcase for saying that, because most modern ecologists think that the challenge of conservation is to keep plants and animals happy by protecting the environment that they are in. They rarely ask the question: could it be somewhere else? They don't have the evidence for how it *might* be somewhere else. That's where the palaeontologist comes in.'

Archer is not alone in proposing species translocation as a means to mitigate the impacts of climate change. However, in reaching back millions of years into prehistory to find an alternative habitat, he is unique. Conservationists have been wary of reaching back any further than the Holocene, lest they be accused of inauthenticity. Yet, with the dividing line between the Holocene and the deeper past becoming increasingly arbitrary from a climate perspective – if the world is on course, as scientists suggest, to resemble the Eocene or the Pliocene – then need we restrict ourselves to preserving this narrow slice of time?

Archer is clear where he stands on such boundaries. 'It's an artificial constraint. It's absurd,' he says. 'Who cares whether you're talking about the Holocene, Pliocene, Miocene? It's worth a try. You always have to think about the alternative to not trialling these things, which is to say: "Sorry, we don't know how to help you. It's been fun. Bye bye." I'm not built like that. And I don't think most people are.'

And so, *Burramys parvus*, with the help of a devoted team of humans, is heading down the mountain and back into the lowlands where its ancestors scampered so many millions of years ago. In a purpose-built breeding facility, the possums will re-learn the language of the rainforest: what food to eat, the vegetation they will encounter, the animals with which they must coexist. Though it may take some coaxing, Archer is convinced that the vocabulary will be familiar and that the introduction has every chance of success.

For, despite his somewhat venturesome attitude towards polar bears, he has done his due diligence as far as the mountain pygmy possum is concerned. Unlike the lizard of Richard Branson's private island, the existing occupants of the rainforest are unlikely to be too annoyed: the extinction of the *Burramys* left a niche that Archer believes is just waiting to be refilled. In other words, the mountain pygmy possum should slot right in. Meanwhile, previous experiments have shown that marsupials can be taught to be wary of potential predators, which should help it to deal with the new threats – foxes and cats – that have arrived in Australia in the last few centuries, against which the possums will have no pre-evolved defence mechanism (although additional fencing may be necessary while they learn).[32]

Archer is encouraged by the fact that the possums have already shown themselves capable of living in unusual environments. Typically, captive individuals have been kept in temperature-controlled conditions, mimicking their mountain habitat. However, in the 1970s, a pair of researchers, Hans Dimpel and John Calaby, maintained colonies at their homes in Canberra. At first, Dimpel kept his possums in his daughter's bedroom, before moving them to the garage and then the verandah. Calaby also kept his possums in an enclosure in his garage. Despite the decidedly non-alpine conditions, the captives thrived and bred.[33]

The point, says Archer, is that these supposedly fragile mammals are more resilient than anyone had thought possible. Given the challenges that lie ahead for the natural world, the worst thing a conservationist could do would be to hold back from trying new approaches; of holding themselves hostage to the precautionary principle.

'We have taken a fair bit of flak about this proposal. But gradually people are backing off,' he says. 'The world is changing too fast. We've really fucked it up, and if we are really going to keep the biological world alive, we have to try these things, and we have to try them very quickly. Because otherwise we are just going to watch a cascade of extinctions taking things out. And probably, in the end, us with it.'

It should be clear by now that, while the deep past may give reason for hope, it provides no excuse for complacency. When it comes to tackling climate change itself, there can be no justification for heeding the precautionary principle. The science is undeniable, and the consequences of ignoring it unthinkable.

Throughout history, global warming has forged world after world: animals and plants merely chess pieces in a 66-million-year game of atmospheric chemistry and cosmic forcing. With humans absent, these upheavals did not matter. There was no one to worry about the growth of palm trees at the poles, no farmers to stress about the arrival of new pests, no cities built upon the coastlines swamped by rising seas. Such changes were neutral and natural – the individual chapters in the story of our planet.

But now we have reached the endgame. The upheavals we are witnessing today are neither neutral nor natural. Humans are to blame, and humans will suffer.

No one is claiming that no plant or animal will be wiped out in the decades to come. Global warming alone may not have caused widespread extinctions – but alone is the operative word in that sentence. Just as the byproducts of volcanic eruptions multiplied the pressures of climate change during the mass extinction events of the past, so modern society has constructed its own stressors. Species face barriers of concrete and tarmac as they seek out cooler climes. Populations decimated by guns and deforestation become genetically weakened, and therefore less able to adapt physically to their new conditions. Refugia are sown with monoculture crops, diminishing any chance of recovery. The more we build over and exploit the natural world, the more these processes of adaptation, through which so much endured for so long, are stopped in their tracks. Meanwhile, the rate of climate change, if not the magnitude, is already outpacing the past. Ecosystems may have the capacity to adapt, but will they have the time?

Ultimately, we do not have the luxury of seeing climate change in terms of flora and fauna alone. The disruption to ecosystems no longer takes place in a vacuum: the impacts ricochet through human society, causing conflict, starvation, disease and general misery. Heatwaves, flooding and sea level rise are already causing death and displacement, with the poorest and most vulnerable often the worst affected.

When it comes to restoring the landscape, we can look to the past for inspiration: for what trees to plant, what animals to reintroduce, what ecosystem to target, and so on. By looking even further back in time, however, we uncover a deeper message: that change is strength. Though humans may bear responsibility for the current upheaval, resisting it is not the answer: we must tackle the root cause. By holding the landscape in stasis – stopping migrations, blocking expansion, refusing to make space for wildlife as it readjusts – we prevent the natural world from doing what it has always done to survive. Indeed, as the world gets hotter, we may find that change is the only feature of the past that we are able to truly preserve.

EIGHT

Redemption

Despite everything – the extinctions, the invasive species, the destruction of the wildwood – there is still the sense that humanity can redeem itself. All we must do is disappear from the face of the earth.

In our absence, nature will recover – or so the narrative goes. The forests will regrow, and rivers resume their natural paths through the landscape. Populations of wild animals will rebound and repopulate their vacant strongholds. There will be carnivores again, fear of which will shape the movements of their prey. Cities will crumble beneath ivy and moss, oil rigs collapse into the sea. The world will rewild. 'Nature is healing,' people proclaimed during the height of the coronavirus pandemic, as dolphins appeared in the canals of Venice and herds of deer roamed the Japanese subway. 'Humans are the virus.'

The preceding chapters of this book have explored the challenges, both socially and ecologically, of returning any ecosystem to a truly natural baseline. The impact of humans has been too extensive, too permanent, for a complete reversal. Since the loss of the megafauna onwards, the landscape has been our creation, its essence moulded by the people who wielded the spears – and our interference has only intensified with time. Destruction, however, is not a one-way street. Forests do regrow. Species can spread. It is not unreasonable to think that, if the original harm was small and distant enough, certain scars may eventually scab over; that the imprint of humanity may be swallowed up as vegetation returns, animals recolonise and nature rebuilds.

But the earth has a memory. Past harms haunt us for longer than ever thought possible. These memories are not friendly ghosts but restless poltergeists – and the forests especially are full of them.

Thuilley-aux-Groseilles is an undistinguished commune in north-east France, surrounded on three sides by a woodland of oak and beech. The land has officially belonged to the village since it was transferred during the French Revolution in 1789. Before that, it was owned by the Abbey of Saint-Mansuy, which was established by Benedictines during the tenth century. The first accurate local map, from 1730, shows the same woodland that stands today, meaning it must have existed unchanged since that point. But probably it dates back a lot further. There are no records to suggest that the land has been cultivated since the Middle Ages; more likely the trees were coppiced to provide wood products for the local monks. In other words, this woodland has been around for a long time.

But not forever. Beneath the groundcover of moss and leaves, the forest bears the remains of a small Roman community. Not much remains beyond the remnants of a few stone walls. However, excavations have shown that the settlement would have consisted of a few houses, tucked to the east of an old road. These dwellings were surrounded by enclosures for livestock: sheep, goats, horses and cattle. Beyond that were terraces for crops, followed by pasture for grazing, and only then by undisturbed forest. Coins and ceramics recovered from among the collapsed walls show that the settlement was established around 50 CE and abandoned some 200 years later. The lack of any subsequent artefacts suggests that the forest moved in swiftly once the people left, and has remained standing ever since.

It was these remains that brought Jean-Luc Dupouey to this particular forest one summer's day in 1999. An ecologist at the French National Research Institute for Agriculture, Food and the Environment, Dupouey was investigating the extent to which present-day forests are impacted by historic land use. He wanted to know how long it takes for a secondary woodland – one that has regrown on land once used for another purpose – to recover to the extent that it becomes indistinguishable from wildwood; to quantify, in other words, how long it takes for nature to heal. 'We wanted to know the length of the legacy,' he told me, when we spoke over the phone. 'What is the duration of the memory of forest ecosystems?'

Scholarship from England had found that these memories could be surprisingly enduring. A study of woodlands in Lincolnshire had found that secondary woodlands were 'conspicuously poor in species' compared to those that bore no indication of human occupation. One of these secondary woodlands was around 400 years old, and still had not caught up in terms of biodiversity. Time alone could not bridge the gap. 'Ancient woods are indeed richer in species than recent woods,' wrote the authors. 'Claims that secondary woods will one day become as rich as ancient, presumed primary woods seem unfounded.'[34]

Still, more than a millennium stood between the oldest forest in that study and the one that Dupouey had found in Thuilley-aux-Groseilles. It had been almost 2,000 years since the first saplings had pushed through the soil of the abandoned enclosures; since the forest had started to reclaim the pastureland for its own. Ample time, surely, for the soil and flora to recover – for the forest to appear, in all respects, as though there had never been a farm there at all. So convinced was Dupouey that all traces of the past would have vanished from the site that he devoted a minimum of time and money to the project, recruiting only two young interns to help with the fieldwork. 'We were cautious,' he recalled. 'I had to recognise that the idea was really strange, even for us.'

The farmers of this ancient settlement would have had to fertilise the soil to ensure their crops returned year after year. The Roman agricultural writer Columella gives a sense of how they would have gone about this task. Some techniques, such as growing cover crops and spreading animal manure and ash, would not be out of place on an organic smallholding today. Others, such as applying human urine to vines to improve the bouquet of the wine, have fallen out of vogue. Either way, the fertilisers used were natural and biodegradable: it seemed likely that, once the humans had vanished and cultivation ceased, the soil would revert to its original state: that the earth would forget the crops it once held, and the old farm would be recolonised by the full suite of woodland flora.

To test the theory, Dupouey and his colleagues marked out dozens of plots across the site, throughout the occupied areas and into the untouched forest beyond. The idea was to compare the flora of the secondary forest to that of the presumed wildwood. Within each one, they counted the flowers and identified the mosses, sieved and sampled the topsoils, and collected leaves from the crowns of the oaks. What they found, to their astonishment, was the spirit of a working Roman farm.

Technically, the formerly inhabited areas of the settlement were more diverse than the undisturbed woodland beyond its peripheries. The sites of the houses, enclosures and terraces were marked out by a throng of common and aggressive species – buttercups, cuckoo flower and dandelion, among others – more typical of meadow than forest. Beyond the embankments, however, the flora had retained its distinctive woodland identity. The pale stems of bird's-nest orchids clustered around the beech trees. Another orchid, the broad-leaved helleborine, grew on a third of the undisturbed plots yet none of the cultivated ones. There was barren strawberry, wild asparagus and mountain sedge. The flora of these remote plots made for a less raucous sight, perhaps, than the carpet of colour that had invaded the enriched soils of the houses and enclosures. But it was certainly a rarer – and from a conservationists' perspective – a more precious one.

A few years later, a group of entomologists returned to Thuilley-aux-Groseilles. They set out insect traps across the settlement, studying the variation in everything from beetles to dragonflies to butterflies to harvestmen. Once again, they found an ecosystem shaped by its former inhabitants. The occupied parts of the site were dominated by insects that live upon the bodies and excretions of other animals: parasites, dung-eaters, blood-suckers. The ancient woodland, meanwhile, abounded with species that depend on plant and wood material for sustenance.[1] It was not just the flora that had failed to recover from past land-use: the biodiversity of the present-day forest was haunted from the understorey upwards. Two millennia later, and nature was far from healed.

Age matters. The complexity of a forest builds across centuries. With time comes death, with which comes life. Ancient trees, for instance, are essential for the survival of certain fungi, which decompose the heartwood and create the hollows that are so valuable for many birds and insects. Rare lichens cleave to their bark, which grows more alkaline with age. Spiders weave their webs in loose fissures, and larvae inhabit the tunnels and chambers of rotting deadwood. With time also comes the increased likelihood of chance events, like storms and fires, that carve out unusual niches within the wider habitat. Its passing brings the arrival of a growing assemblage of specialised flowers; species that, for one reason or another, struggle to spread into new areas – perhaps their seeds are dispersed only by ants, or they produce no seeds at all, instead reproducing clonally, creeping outwards by centimetres with each turn around the sun. In a forest, time creates an effect as impressive as any gothic cathedral: the understory as luminous as a stained glass window, the canopy as imposing as any flying buttress.

But age is not everything. Just as important is continuity: whether an area of forest has always been forest, or whether it has regrown on the ruins of human settlement – an old Roman farm, for instance.[2] A primary forest is one that has been standing uninterrupted since prehistory. But to distinguish it from secondary forest, one must do more than count the rings within the trunks. A forest is like a human community: the time of its establishment cannot be judged by the age of its current residents. Just as a secondary forest may contain trees that are hundreds of years old, so a primary forest may feature young saplings: the next generation growing up alongside the elders. A primary forest may even have been subject to human management, such as coppicing or pollarding. The only constant is the forest itself. It is this longevity that imbues it with its unique richness, for time alone cannot compensate for the fracture caused by clearance, and, in particular, cultivation.

The reason for this lies not in the above-ground architecture but down below: in the dirt. Agriculture changes the soil in a fundamental way. When farmers apply fertilisers, returning the nutrients that they have taken out, they change the earth irreversibly, regardless of whether that fertiliser comes out of a cow's behind or an industrial manufacturing plant. One of these nutrients is phosphorus, an element that is extremely stable when fixed in the soil. An ancient forest acts as a closed-loop system, the nutrients absorbed by the trees each spring only to be returned to the soil come autumn when the leaves fall to the ground.

Phosphorus is everywhere. It is found in the earth's crust, but also in urine, bone ash and guano, for example. Archaeologists use it to trace human activity: its presence has identified everything from Aztec sacrifice rites to the decomposed corpse within the burial chamber of the Sutton Hoo.[3] In such instances, the trail of phosphorus was the byproduct of death. For farmers, however, its addition to the soil is intentional and necessary – and occasionally achieved at some cost. In 1890, around 180,000 mummified cats from Egypt were sold at Liverpool Docks, having been discovered after a farmer fell through a hole into a catacomb the previous year. Almost all were crushed and spread over the fields of northern England.[4]

With the addition of phosphorus, crops meant for human consumption grow fast and strong. For the flora that depends on the undisturbed soils of an ancient forest, however, the addition of this element can prove a death sentence. The enriched earth poses an impenetrable barrier to slow-moving species as they attempt to recolonise the understorey. In one experiment, scientists planted oxlips – a classic ancient woodland species – in woodland growing on formerly fertilised farmland.[5] Naturally, the species spreads slowly, but the scientists hoped to speed up the process of colonisation through introducing it by hand. It wasn't enough: compared to those in primary forests, the planted oxlips lived shorter lives, struggled to reproduce and failed to compete with more ruthless species, like brambles and nettles, that boomed on the fertilised earth. After eight years, the scientists were left with fewer than they had planted in the beginning. 'If you determine the conservation value of a forest based on plants, then it is definitely clear that recent forests haven't much value,' said Martin Hermy, one of the ecologists behind the study. The past, it seemed, could not be easily recreated within the conditions of the modern world.

At Thuilley-aux-Groseilles, Dupouey found that phosphorus levels were highest near the areas of human habitation. That phosphorus not only affected the flora of the understorey at those sites, but also the oak trees. Those that had regrown where the houses and enclosures had once stood were large and luxurious compared to those within the undisturbed areas, their development spurred on by nutrients applied almost two millennia in the past. By spreading ash and manure, and who knows what else, the farmers had transformed the earth and, with it, the forest. The impacts of fertilisation had outlasted the people, the animals, the settlement, even the Roman Empire itself.

The results were astonishing, but that was one farm in one forest: not enough to draw any wider conclusions about the endurance of the past. And so, years later, Dupouey repeated the experiment on a much larger scale, in the Tronçais forest in central France. This is a place famous for its high-quality oaks, which today are harvested to make barrels for vintage wines; in previous centuries, the trees were coppiced for charcoal and clogs. Old maps showed that the forest had stood there, its boundaries unchanged, for centuries. But Tronçais, despite appearances, was not a remnant of the wildwood. Around two decades ago, archaeologists discovered the remains of more than a hundred Roman settlements, including two temples, interred beneath the earth and grass. It was upon these ruins that the forest had regrown.

Just as they had done at Thuilley-aux-Groseilles, Dupouey and his colleagues set up a meticulous inventory of every aspect of the forest – and once again they found a landscape shaped by its long-dead inhabitants.[6] The plants that clustered in the centre of the settlement were those that thrived on neutral and enriched soils, while ancient forest species preferred the remoter areas beyond the remains. A separate study found that even the ectomycorrhizae – fungi that form a symbiotic relationship with plant roots – were affected by the nutrients, with communities of species determined by whether the site was occupied or undisturbed in the distant past.[7] This time, there was enough evidence for Dupouey to feel certain about what he had found: forests have a memory, and that memory is longer than we know.

'We want our forests to be resilient. We think they are resilient because the trees are always coming back. You have to be very efficient to remove them forever,' he reflects. 'But if you look under the trees, you see that it is no longer the forest we had before. We had an impact. And that means that forests are not as resilient as we think.'

The word 'ancient' conjures images of classical empires and fallen civilisations. Those images, however, do not apply to the woodlands we call ancient today. Most are not remnants of the primeval wildwood but something much more recent. Given that it is generally impossible to determine the age of an old forest, the threshold between ancient and modern is instead set at the point when a country developed its first reliable maps. In England, the line is drawn at 1600 CE, in recognition of the Elizabethan respect for cartographical accuracy following the somewhat fantastical efforts of their medieval predecessors. In

Scotland, it is set at 1750 CE, following the military survey that led to the creation of the Roy Maps. The cut-off point in France is later again, based upon the production of the Napoleonic cadastre in the early nineteenth century. Historians previously supposed that, if a land had been forested at these points, then it must have always been forested. Technological developments, however, are increasingly proving that this is not the case.

Woodland has always waxed and waned across Europe, its expansion generally tracking the course of human misery. The collapse of both the Greek and Roman empires was followed by periods of reafforestation as populations declined and farms were abandoned.[8] The Black Death, which killed off millions of peasants across Europe during the fourteenth century, resulted in the mass rewilding of arable land.[9] The murder, starvation and death by disease of up to 90 per cent of the Indigenous population in the Americas, following the arrival of Columbus in 1492, resulted in so many new forests that the additional carbon sequestration has been credited with causing the Little Ice Age.[10] The ancient woodlands of today, in other words, are not the remnants of wildwood we once supposed them to be. They are the overgrown graveyards of human civilisations, whose traumas are remembered by the soil long after they have faded from our minds. Ecology bears witness to the past.

In recent decades, the use of airborne lasers has allowed researchers to peer beneath the canopy at an unprecedented scale, unveiling the mass of human debris scattered throughout even the most supposedly pristine ecosystems. Such investigations have revealed a landscape that is more haunted than we ever imagined. In fact, the remains of Roman farms are being discovered at such a rate that researchers are beginning to question whether perhaps ancient deforestation exceeded even modern-day efforts. 'It's a strange story – nearly everywhere we find them,' says Dupouey. 'And if you have everywhere farms dating from antiquity, then the question is: where were the forests?'

Take the Białowieża Forest: Europe's primeval wilderness *par excellence*. It is where, in chapter six, we met the last of the continent's free-ranging bison, who were able to take refuge amid its ancient trees even as they succumbed to human pressures elsewhere. The forest has never been entirely free from anthropogenic influence. In prehistoric times, it was a source of meat, mushrooms, berries and honey for hunter-gatherers. But, until recently, the only known physical evidence for human activity was a small number of graves and cemeteries, dating from between the eighth and twelfth centuries, rising from beneath the moss-covered ground: remains more associated with ritual than cultivation. In 2016, however, researchers discovered evidence for widespread farming within its interior: over 300 km of field boundaries and plough headlands, constructed by Romans between the first and fourth centuries CE. Once again, the soil remembered their presence, long after their bones and buildings had crumbled, with samples from the once-arable land showing significant changes in the levels of carbon and nitrogen present.[11]

Fine: we already know that Europe's early inhabitants had a cataclysmic impact on the forests. That we should still be finding evidence for their presence is unsurprising, even if the longevity of their impact upon the vegetation is bewildering. But what about the High Arctic, a place that supposedly remained pristine until the arrival of the Europeans? The Canadian archipelago was occupied by Thule whalers – the ancestors of modern-day Inuit – from around 1200 CE. It has generally been assumed that they had a negligible impact on the local environment: their societies were semi-nomadic and widely dispersed, and their food supplied by hunting and gathering, no cultivation required. Even the houses of their winter camps were made from natural materials; namely, the massive bones of the bowhead whales that they harpooned for food and other resources during the summers.

Yet even these people, who lived so lightly upon the land, left a footprint that can still be detected today. Thule settlements are easily identified from the air: not only by the bleached bones that stand out against the tundra, but by bursts of lush vegetation – mosses, saxifrage and willow – that cluster around the remains of the whales. These plants are the beneficiaries of nutrients leaching from the bones into the soil.[12] The ponds adjacent to the villages, fertilised by centuries of decay, still contain elevated levels of phosphorus and calcium, and unusual assemblages of algae, compared to unaffected ponds. Scientists believe that recent climate change may even be causing these nutrient inputs to spike as warmer temperatures accelerate the decomposition of the remaining bones – an unnerving call-and-response echoing across time between two very different civilisations. 'It is ironic,' they write, 'that the High Arctic, generally considered to be the last refuge from local human disturbances, contains the oldest record thus far obtained in the United States or Canada of a human population affecting freshwater ecology.'[13]

None of that is to say that these ecosystems are ruined. But they are different – unnatural – in ways that we can now barely recognise, so accustomed have we become to a world shaped by humans. By observing how past land use has influenced modern ecology, we can make better decisions around how to conserve and restore our forests and lakes. But perhaps more importantly we are forced to assess our own behaviour. These ghosts carry a warning: time does not heal all wounds. Even shallow cuts can fester.

What kind of ancestors will we become? Industrial farms and sprawling megacities have nothing in common with the crooked fields of Roman France or the whalebone houses of the Thule. We live more wastefully, more excessively, more intensively than at any time in the past. We drench our crops in agrochemicals, burn coal and bury nuclear waste, depend upon materials that take centuries to decompose. Our descendants will not need to count oxlips in ancient forests or sample the mud of Arctic lakes to uncover our environmental legacy. It will be written in bold – in blood – across the surface of the earth: in the scree that once held glaciers, the plastic on the remotest shores, the silence where there should be song. The earth has always remembered. Now, the human race is making itself more unforgettable than ever before.

Yet there is hope amid the gloom. The earth remembers the good times, too. We have seen, in preceding chapters, the myriad ways in which the past has left its imprint upon the present: fossils pressed into stone, pollen preserved in the mud, the adaptations still borne by animals and plants to creatures now extinct. Most of these memories are inert; they exert no influence upon the ecosystems of today. But not all of them. For all the malevolent poltergeists that haunt the natural world, there is the occasional friendly ghost. It was in pursuit of one such spirit that I ended up in a field in Norfolk, early one morning, on what should have been the first day of my honeymoon.

My new husband dropped me on the corner by an old flint church, before driving off to the nearest village for several cups of coffee. I was bleary and disorientated from the whirlwind of the past week, and the feeling of being suddenly alone by a quiet graveyard, draped in fog, felt almost uncanny. I had been given a grid reference and told to follow the track, which would lead me to the field where the séance-of-sorts was underway. However, as soon as I turned the corner, I was greeted by the sight of a JCB looming up from behind a hedgerow, taking bites out of the earth like a big yellow dinosaur. This must be the place, I thought.

The field itself was unassuming. After centuries in cultivation, the present owner had decided to let it rewild. The ground had been taken over by long grass, punctuated by the brown stems of dead dock leaves and bursts of yellow ragwort. Old maps showed the presence of two ponds, but these had vanished by the 1880s, filled in like so many others across England and Wales. During the nineteenth century, the countryside was dimpled by around 800,000 such ponds, which were tightly interwoven with rural life. They were dug to extract marl, a mineral-rich material used to improve the soil, and to provide drinking water to the shire-horses that ploughed the fields. Their presence was so valuable that farmers would remove the sediment and cut back any scrub that threatened to overwhelm them. Though their purpose may have been pragmatic, these watering holes also acted as islands of aquatic biodiversity. Ponds are among our most humble habitats, ignored or even reviled – 'pond life' is hardly a compliment – yet together they support two-thirds of freshwater species.[14]

A few centuries ago, the house adjoining this particular field was occupied by a man named Theophilis Girdlestone, the rector of the neighbouring parish. Should the reverend have ever felt like shedding his cassock and taking a dip in the quiet seclusion of these ponds, he may have unwittingly crushed a wandering pond snail with his toes, or dislodged a pea mussel as his feet squelched into the soft mud at the bottom. Perhaps he would have felt the lord's presence in the pond skaters as they walked across the water – or the devil's work in the water boatmen, an insect with a famously painful bite, and the leeches that may have latched onto his legs. But maybe the beauty of the Creation made up for it: the metallic gleam of the dancing dragonflies, the irises with petal-mouths agape, the water-crowfoot waltzing round him as he swam.

But neither beauty nor biodiversity could save them – nor the many others like them across the country. During the twentieth century, the majority of the UK's ponds were lost. Many were intentionally filled in following World War II, when the intensification of food production meant that farmers were encouraged to grow crops on every last inch of land. Others were neglected and eventually buried beneath the encroaching vegetation. By the 1980s, just a quarter remained. Yet, like the Roman farmers of Thuilley-aux-Groseilles, these ponds left an indelible impression on the landscape. If you have ever noticed mist pooling over a certain patch of land, or found a puddle that refuses to drain, or observed a strange pattern in the crops, then it is possible that you are in the presence of one of these 'ghost' ponds.

'Once you know the signs, you'll see them everywhere,' as Carl Sayer put it to me that morning in the Norfolk field. A professor of geography at University College London, he has pioneered both the study and the restoration of these lost habitats. While his academic career has focused on aquatic conservation at large (he has published papers on everything from carp decline to otter diets) on this occasion the job was personal. He was born in this village, spending his boyhood catching fish and trapping newts in the ponds that still existed at that time. His dream, he told me, was to leave the landscape better than he found it. Today, that dream was moving one step closer to reality.

It was Carl who had invited me to this field to watch the resurrection of a ghost pond in action. Though it was not his first time overseeing such a project – around 25 have been brought back from the dead in this way – he still harboured an undisguised enthusiasm for what was afoot. He stood at the edge of the deepening hole, his jumper flecked with pale seeds from the surrounding field, watching as the JCB peeled back the layers of crumbly, coppery topsoil. He was waiting for something specific: the dark layer of sediment that would indicate they had reached the bottom of the old pond. Sometimes you can even smell it, he told me – the muggy aroma of algae, frogspawn and summer afternoons.

Uncovering this rich stratum is the goal of the whole operation. Because this is where you will find them: the seeds, the eggs and the oospores that have so long lain dormant in the soil.[15] For a century and a half, these propagules, released when the pond was at its peak, have been hibernating beneath the earth, in the darkness, waiting like bulbs for their own personal spring. Given the opportunity, the seeds and oospores will germinate and grow, the eggs hatch into invertebrates ready to helicopter through the water. Carl is providing that opportunity. By removing the upper layers of mud, and waiting for the hole to fill with rain, these propagules will have everything they need to live another day. 'It's a really exciting way of bringing ponds that have been completely lost back to life – because they were never dead,' he said.

Almost two metres below the surface, it appeared: the mud of the old pond. I took a pinch and rubbed it between my fingers. It was as smooth and unremarkable as wet clay. The life latent within was invisible to the naked eye. To see the soul reveal itself, I would have to wait a short while longer.

Nature is fragile but also tenacious. Ecosystems are easy to destroy, but difficult to obliterate in their entirety. This can work in our favour. It means that apparent disappearances, even extinctions, are not as final as they might appear; that we must not always be starting from scratch. This is true not only for ponds but also forests. A few years ago, in Exmoor National Park, rangers were astonished to see a carpet of bluebells – an indicator of ancient woodland – emerge across a bare hilltop. They had been clearing bracken to prepare the ground for planting new trees, and the disturbance seems to have reactivated the bulbs that had lain dormant for centuries beneath the soil, ever since the original woodland was felled.

Infilled ponds are particularly suited to preservation. Seedbanks survive well in dark, wet, oxygen-deprived conditions. Pond mud has all three of these qualities. Charles Darwin was among the first to discover how much life lays latent in sediment. In *The Origin of Species*, he described how, one February, he took three tablespoons of mud from the edge of a little pond. He kept it covered in his study for six months, pulling up the shoots as they emerged: 'the plants were of many kinds, and were altogether 537 in number,' he wrote, 'and yet the viscid mud was all contained in a breakfast cup!'

The mud of Darwin's experiment had the advantage of coming from a real pond, still flush with life. He illustrated the astonishing number of seeds that will sprout from a small amount of mud, but not the longevity of those seeds. More recent research, however, has shown that, in extreme environments, a seed can remain viable for millennia. In 1995, scientists managed to grow the seed of a Sacred Lotus that had been immersed in a lake bed in China for almost 1,300 years – a relic of the crops cultivated by some of the first Buddhists in the region.[16] More recently, Russian biologists succeeded in regenerating a 32,000-year-old plant – narrow-leafed campion – from fruit tissue buried by squirrels in the permafrost of Siberia.[17] Scientists have even succeeded in germinating herbarium specimens, kept in dry storage for up to 150 years, raising hopes for the revival of species that have since gone extinct in the wild.[18]

This ability to bed down and await suitable conditions is known as 'temporal dispersal'. Rather than spreading from one place to another, caught on the wind or the fur of a passing animal, some seeds are able to spread through time, spawning new generations separated by hundreds, even thousands, of years.

So far, Carl and his colleagues have shown that it is possible to germinate species from seeds and oospores that have been buried in pond mud for at least 150 years.[19] Although a century or two of hibernation may sound unimpressive in comparison with the revival of an Ice Age campion, it is worth remembering the conditions in which each was preserved: one, beneath 38 metres of undisturbed permafrost; the other, in a field that had been intensively farmed for decades, the soil compacted and drenched in chemicals. Then there are the interventions required for their resurrection. The development of the Pleistocene plant tissue was induced in vitro, and then nurtured in light- and temperature-controlled conditions. The seeds from the ghost pond, meanwhile, required no special treatment, no laboratory equipment, to bring them back to life. The process is quick and cheap, and the opportunities are many. That these ecosystems can rise from the dead, irrespective of the sins heaped upon them, seems nothing short of a miracle. You cannot help but feel that Reverend Girdlestone would approve.

The restoration of ghost ponds also brings with it the possibility of surprises. A few years ago, grass-poly appeared around the edges of an old cattle-watering pond that Carl and his team had recently brought back to life. The flower, now endangered, had not been seen in Norfolk for more than a century.[20] Its emergence demonstrated the advantage of resurrecting forgotten pools over simply digging new ones. The latter approach may seem simpler: there is no need to consult old maps or watch for pools of mist in the fields. But it depends upon species colonising from neighbouring land. In a countryside bereft of life – and, in particular, of aquatic life – this can lead to a disappointing turnout. Ghost ponds, meanwhile, act as a portal into a world when farming was less intensive; when wildflowers and animals could still thrive alongside crops and grasses. 'They were dug at a time when the landscape was a lot wilder, more connected, pre-fertilisers,' said Carl. 'When the countryside was more vibrant and the ponds were more vibrant.'

I didn't spend more than a few hours at the ghost pond. Once the JCB had done its work, there wasn't much to see. But, as the year went on, I sometimes thought of that hole: whether the rain had filled it with water, and if anything noteworthy had emerged from the mud. When I felt enough time had passed, I sent Carl a message, asking for an update. I needn't have waited so long. He sent back a series of photographs, showing it filling up over a period of just ten months: a small brown puddle rising, rising, rising, over the layers of mud and topsoil, gradually turning into a deep green pool. There were clouds reflected on its surface, and billows of weed down below, working their way towards the light. It had become home to fifteen wetland plants, including fine-leaved water dropwort – the only example of this species in the local river catchment. I wrote back to him: would that dropwort have been one of the plants in the original nineteenth century pond? He replied: '100 per cent.'

In ecology, the idea of resurrection is controversial. Headlines cluster around efforts to bring back the big-name beasts: mammoths, dodos and thylacines. Whether this is a worthy goal or a flashy waste of money depends on your perspective; the topic has been debated to death in the pages of both popular magazines and scientific journals. The concept is undoubtedly captivating: a chance, however slim, at seeing the most charismatic elements of the natural world brought back to life – creatures whose mystique has been enhanced, no doubt, by the fact that we can only imagine how they looked and behaved when they were alive in the wild.

The resurrection of ghost ponds seems a humble task by comparison. We are talking about the return of habitats associated not with untrammelled wilderness, but of village greens and weedy fields. No one pines for pondweed, and even the most ardent conservationist would struggle to romanticise the return of the water flea. Mostly no one has heard of the rare flowers that sometimes show up. To invest our hopes of redemption in such a habitat almost feels like an acknowledgement of how far we have drifted: of how small and tame our world has become.

But there is something about these ghost ponds that sticks in my mind. Because the restoration of these gentle habitats is not creating a facsimile of the historic landscape. It is not planting trees in imitation of the original wildwood, nor reintroducing livestock as proxies for the lost megafauna. It is scratching back the surface to reveal the real thing. It is time travel.

The thought makes me nostalgic for a world I have never known. I want to lie back in long grass, inhale the dewy smell of pondweed, and dip my toes in the cool water, like a vicar who thinks no one is watching. I want, just for a moment, to not feel like even the humblest of habitats have been ruined by humankind. To believe that, despite millennia of destruction, redemption is still possible. That such a thing may come to pass, right here in this Norfolk field, seems nothing short of a blessing. It wasn't such a bad start to my honeymoon.

NINE

Hallowed Ground

For the San Carlos Apache Tribe of Arizona, Chi'chil Biłdagoteel is a sacred place. The name translates loosely into English as Oak Flat; a reference to the groves of Emory oaks that grow there, which produce acorns so uncharacteristically sweet they can be eaten straight from the bough.

Officially, this cactus-covered plain belongs to the American government. The tribe was exiled in the nineteenth century and confined to a small reservation to the east, dubbed Hell's 40 Acres due to its barren landscape. Politicians never honoured a subsequent promise to return Oak Flat to the Apache. Instead its people were massacred by soldiers and civilians looking to profit from their land.

But although the colonists could expel the tribe, they could not banish the gods. The Apache have long held that this desert plain offers a direct line to the Creator. Amid its jagged cliffs and grassy basins dwell the Ga'an – angel-like beings that act as a buffer between heaven and earth, offering guidance to those who seek them.[1] It is here that the tribe gathers to carry out its most important rituals, and where its people forage for medicinal plants that cannot be found anywhere else. Just as a baby cannot be baptised in a kitchen sink, so these customs cannot be carried out on another patch of land: its divinity is non-transferable. 'I call Oak Flat,' says Terry Rambler, chairman of the tribe, 'the Sistine Chapel of Apache religion.'

The trouble with divinity, however, is that it also cannot be quantified. In 2014, the government passed legislation authorising yet another transferral of the ownership of Oak Flat. The beneficiary this time was Resolution Copper, a subsidiary of the mining giants Rio Tinto and BHP. What made the land precious, in the company's eyes, was not its spirits but the deposits of copper, buried within the bedrock, over a mile beneath the ground. Copper that could be brought to the surface and sold.

Extracting the ore would mean tunnelling beneath the seams, exploding it into chunks and carting it to the surface. This, eventually, would cause the land to subside and an enormous crater to sink into the landscape. By the company's own admission, the procedure would cause irreparable damage to the natural and cultural environment, wiping out rare species, drying up sacred springs, disturbing old burial sites, and destroying the ancient groves that lend the site its name.[2] 'The proposed destruction of Oak Flat is akin to tearing out pages from a history book in the library,' wrote one researcher, in testimony to Congress.[3]

The Apache, once again, are fighting for their land – and for their gods. Their battle, however, is just one front in a wider war to protect the earth's sacred sites. I could have equally written here of the Hawaiian elders arrested for obstructing the construction of a telescope on the summit of Maunakea, considered the first-born mountain son of Wākea and Papa, the creators of earth and sky. Or told the story of the Djab Wurrung women of Australia, who fought to save their ancient birthing trees, beneath which placentas are buried, from being cut down to make way for a new highway. For so many people across the world, the destruction of the land is more than vandalism: it is desecration.

Sacred land has been called a 'shadow conservation network' because of the rich ecosystems that have emerged in the absence of human exploitation. Oak Flat, for instance, is home to several threatened species, including the Arizona hedgehog cactus, yellow-billed cuckoo and narrow-headed garter snake. Nature thrives in the presence of the gods – but only so long as it remains in the hands of those who revere them.

The concept that the land itself can be sacred is more or less absent from Christianity. In the Book of Genesis, paradise is a garden – Eden – where nature is tended by man. The wilderness beyond, into which Adam and Eve are banished after eating the forbidden fruit, is hostile and cursed. These days, its holy spaces are built as any other: from stone and mortar. The Sistine Chapel of the Christian religion, after all, is the Sistine Chapel.

'To a Christian a tree can be no more than a physical fact,' wrote the historian Lynn White Jr in 1967 in his now-classic essay, *The Historical Roots of our Ecological Crisis*. 'The whole concept of the sacred grove is alien to Christianity and to the ethos of the West. For nearly two millennia Christian missionaries have been chopping down sacred groves, which are idolatrous because they assume spirit in nature.'

That is not to say that Christianity is entirely without its holy ecosystems. In the mountains of Greece, networks of sacred groves have sprung up around rural villages, standing guard over outlying churches or monuments dedicated to the Virgin Mary and the saints.[4] Studies have found that these groves are richer in fungi and songbirds than nearby unconsecrated land.[5] Similarly, the brittle farmlands of Ethiopia are studded by lush forests, emanating outwards from the buildings of the Orthodox Tewahedo Church. These are all that remain of the native woodlands that once clothed the north of the country, preserved in these small fragments as a stand-in for paradise or as homes for hermits.[6] Closer to home, I am often struck by the quiet beauty of England's ruined monasteries – where divinity seems to persist in the harebells that hang from cracked ledges and the ferns that unfurl behind the gravestones – and by the botanical surprises that emerge from the unploughed soils that surround so many village churches.

These ecosystems, however, have never really been considered divine in their own right. The presence of God is spillover from the manmade structures that they accompany; the churches and shrines constructed, stone by stone, by human hand. Indeed, the chapels that smatter the Greek mountains are considered an effort to tame the wilderness, the groves that surround them a buffer against the demons that dwell beyond the periphery.[7] In Christianity, fragments of bone and scraps of cloth inspire a level of devotion that is rarely invoked by the natural world. The prevailing attitude is more often one of suspicion – as though the worship of yew trees and springs might somehow pave the way to paganism – than of reverence.

Places like Oak Flat will always have their defenders, regardless of the species that exist there or the ecosystem services they provide. It is the presence of the gods, the spirits, the ancestors, that justifies the cause. In England, by contrast, the arguments for saving a site often rest upon the back of the great-crested newt.[8] It is not hard to see which inspires the greater passion.

Did such reverence for nature ever exist? Britain does have one holy standard-bearer for the natural world: Saint Cuthbert.

Today, Cuthbert is associated with the Holy Island of Lindisfarne, where he was bishop during the seventh century. The island, just off the coast of Northumberland, remains a place where spirituality and ecology overlap. When the tide is out, it can be reached on foot. I am among the many people to have slipped off my shoes at the start of the causeway and made the journey across the mudflats. There is an immaculate quality to the landscape that makes you want to immerse yourself in its sulphurous depths. The sensation of samphire and seagrass squelching between your toes is strangely spiritual; a baptism of mud and salt.

But again, it is stone and mortar that lends Holy Island its holiness. The medieval priory, founded here by monks in 635 CE, is what draws most pilgrims across the sand. Travel a few nautical miles up the coast, however, and you will find the Farne Islands rising from the North Sea. They are wild and uninhabited now, just as they were in Cuthbert's time, save for Cuthbert himself. For many years, the saint lived as a hermit on Inner Farne, the biggest of the islands, passing his days in a small cell of stone and turf.

For it was the Farne Islands, with their sheer cliffs and raucous colonies of seabirds, that the monks considered to be inherently sacred.[9] While Cuthbert was said to have made the island spiritually habitable, banishing its devils and creating a natural spring, they believed that he was only tending to a quality that already existed in the land itself. 'For this place is holy, which before my coming was not cultivated in the proper reverence,' said Bartholomew, a twelfth-century hermit, justifying his wish to be buried on the island. Ordinary people also understood its sanctity. One story, recorded by an anonymous writer around 1200 CE, tells of a young labourer who was kidnapped by fairies and forced to drink ale from a green horn, which rendered him dumb. His voice only returned after he visited the Farnes and accepted a drink of holy water.[10]

The Farne Islands are the only place that I have encountered that could be described as sacred in its own right – whose holiness lies not in the presence of monks and their monasteries, but in the land itself. There must have been something about its salt-whipped cliffs, whitened with guano and feathers, that suggested this was hallowed ground.

The ancient sanctity of the islands, however, has been all but forgotten in recent times, their light dimmed by the greater glory of Lindisfarne, where the various spiritual retreats on offer provide more obvious opportunities for prayer and contemplation. Journeys to the Farnes are rarely made for religious reasons these days; the regular boat trips instead focus on wildlife and shipwrecks. The only hermits are the National Trust rangers, who spend ten months of the year in isolation, caring for the colonies of puffins and terns that nest in the solitude of the rocks.

What hasn't been forgotten, however, is Cuthbert's affinity for eider ducks. Even today, in the north-east, these birds are commonly known as Cuddy ducks, in reference to the saint himself. In his twelfth century account of the miracles of Cuthbert, a local monk called Reginald describes how the saint had tamed the birds to such an extent that they would approach the tables and nest beneath beds. So strong was the saint's affection for the birds that he forbade anyone from hunting or harassing them on Inner Farne – although they were free to exploit them elsewhere. Those who broke these rules faced various degrees of divine retribution. One unfortunate servant ate one of the birds and attempted to hide his crime by scattering the remains at sea, only for the bones, feathers, beaks and claws to wash back up inside the door of the church. He repented of his sin and was forgiven. Another local man, called Roger, was not so lucky: he disturbed the flocks and died three days later.[11]

As a result, Saint Cuthbert is often said to have established the world's first ever nature reserve. And in some sense he did: environmentalists and politicians continue to invoke Cuthbert's name even today as an argument for protecting the eider ducks, which are now endangered across Europe. As recently as 2018, his ancient sanctions were cited in parliament, by the local Conservative MP, as a reason for protecting those that still swim off the Northumberland coastline.[12]

Yet, as usual with these old stories, there is an element of invention after the fact. The earliest evidence for any affinity between Cuthbert and these birds is his burial robe, nicknamed the Nature Goddess Silk, in reference to the long-haired woman and stylised ducks that it depicts. Historians, however, have argued that these ducks are not eiders but rather an unspecified species drawn in a manner that was typical of Persian art at the time – and, moreover, that the silken robes were probably added to the coffin several centuries after his death. The first explicit written reference to eider ducks does not appear until the twelfth century, some five centuries after his death. The Venerable Bede, Cuthbert's earliest hagiographer, does not mention the birds at all.

As such, some have taken the critical perspective that the eider legends are nothing more than parable – a tale to illustrate the special relationship that existed between the saint and his flock, or Cuthbert's dominion over nature, rather than any particular affection for the actual birds – and argued that romantic notions of nature protection were only projected onto Cuthbert later, with the rise of the conservation movement. 'The notion that the historical St Cuthbert established the world's first bird sanctuary, or indeed had any particular tenderness towards animals, is a modern one,' writes the folklorist Antone Minard.[13]

The truth of Cuddy and his ducks will always be a mystery. Even so, I like to believe that the affection between them was real. Perhaps that is simply wishful thinking, or my sweet spot for the saint after having lived in the north for so long. It is worth noting that the eiders did not have to perform any particular feat of obedience to receive Cuthbert's grace, in contrast to other similar stories. In fact, Reginald gives the impression not of meek servants of the Lord, but creatures that occasionally overwhelmed the monks with their irrepressible wildness: the Farne Islands during the breeding season were no peaceful Eden, but a raucous carnival of feathered debauchery. 'These specially blessed wild animals do not seem to have any precedent in hagiographic tradition,' writes the historian Dominic Alexander, in his masterful book on the topic, *Saints and Animals in the Middle Ages*.

Regardless of Cuthbert's feelings towards them, I cannot bring myself to believe that the eider ducks are genuinely sacred. I do love them – their monochrome markings, the contented look of the females as they squat upon their eggs – but I do not personally consider them to be holy. Nor, I suppose, do those who have invoked Cuthbert in the name of marine conservation. Rather, the reference to the saint is an evocative way of calling attention to a pre-existing cause: a prime example of what the Hebridean scholar, Donald Meek, has termed 'saint seeking'.

'Nowadays, there may be many "saint seekers" who have only the vaguest connection with Christianity in any form,' he writes in his book, *The Quest for Celtic Christianity*. 'Their reasons for seeking the saints may be quite diverse, in many cases unashamedly secular rather than devotedly sacred. As a result, the quest for the saints may be motivated by factors ranging from general curiosity to a special concern for the local economy or the environment.' Saint Columba, an Irish priest who founded the abbey on the Hebridean island of Iona, is another whose image has been reappropriated: one story tells of how he nursed a sick crane back to health after a troubled flight across from Ireland. The act is often portrayed as a sign of his kindness to animals – I have even seen it claimed that he founded the first animal hospital[14] – when his sixth-century hagiographer makes it clear that the saint's concern was actually for the crane as a symbol of the strength of his native country.

So often, the story comes after the cause.[15] We protect land because it is beautiful, or because it hosts a rare species, or because it acts as a buffer against floods, but rarely because it is inherently sacred. Most often, however, we fail to protect the land at all. Genuine reverence rarely draws people to the front-lines of environmental protest these days. Without a higher purpose, the earth lies defenceless.

I had all but decided that conservation and religion were ultimately two separate worlds – that Christianity had no equivalent landscape to Oak Flat – when, quite by accident, I stumbled across the story of Saint Melangell. She almost changed my mind.

St Melangell's Church sits at the head of the Tanat Valley, in North Wales, in a landscape carved out by glaciers. The nearest settlement is Llangynog, a little village whose pretty stone terraces are set off by the black scree slopes that loom down from the north. It is here that most pilgrims will start their journey, leaving their vehicles in the car park before covering the final two miles to the church on foot.

This is no Camino de Santiago. The road is flat, tarmacked and lined by hazel hedgerows, tracking the path of a shallow stream. Occasionally the path sinks into holloways, hinting at the many feet that have walked this way in the past. For centuries, the cult of Melangell drew people from far and wide, seeking healing and succour beside her bones, which lie beneath the large stone shrine inside the church itself.

Saint Melangell is associated, like Cuthbert, with animals – specifically, hares. She was, according to legend, an Irish princess who fled her father's court sometime during the sixth century to escape an unwanted marriage. After crossing the ocean, she wound up in the Tanat Valley, where she lived alone in the wilderness for fifteen years, never once setting eyes upon a man. That changed in 604 CE. The land she occupied belonged to Brochwel, Prince of Powys. He was out hunting with his hounds when he encountered Melangell praying within a thicket of brambles, the hare that he had been pursuing staring back at him from beneath the hem of her cloak. The prince encouraged his dogs to seize the animal, but they backed away, howling. Astonished, Brochwel asked Melangell how long she had lived in such a wilderness, whereupon she recounted her story. Impressed by her devotion, Brochwel donated his land to her as a perpetual sanctuary, where she remained for another thirty-seven years, founding an abbey and continuing to commune with the hares.[16]

How much truth there is in this story is impossible to know. As with Cuthbert, the first account of her life comes well after her death. However, the discovery of a woman's skeleton in 1958, beneath the raised stone marking the spot of Melangell's burial, suggests that there is something to it.

What I found most intriguing about the story of Melangell was that it was the land, rather than the church, that was her sanctuary – the wilderness, rather the building, where hares received divine protection. The story kept its power over time: in the late seventeenth century, the antiquary Thomas Price wrote that 'of late yeares no man would offer to kill a hare in that parish', and that 'when a hare is persued by dogges, if any cry God & S. Monacella bee with thee, shee is sure to escape'. Although he spoke of it as a faded tradition, the concept of the valley as a place of sanctuary has stood the test of time, as the owners of the local estate discovered a few years back.

The Llechweddygarth Estate sits back from the main road: a grand old house that a pilgrim could easily pass by without ever knowing it was there. For as long as anyone can remember, it has hosted a pheasant shoot, with people – mostly men, mostly wealthy, occasionally accompanied by their butlers[17] – arriving in droves to shoot birds in vast quantities, across 5,000 acres of land. In other words: the valley has not been a sanctuary in any meaningful sense for a very long time.

However, it was only recently that the gunmen came to blows with Melangell's devotees. In 2017, the estate went up for sale, and was purchased by a businessman who had made his fortune in the haulage industry. He hired a youthful new manager, Martin Lott, to run the shoot – and that is when the trouble in the valley really began.

With the sale of the estate came plans to increase the shoot. The previous lease had stipulated that shooting could only happen two days per week, but with the purchase of more land from a neighbouring estate there was scope for more days, with almost double the number of locations. I tried to get in touch with Lott to ask him about his plans but received no response.

However, he told the *Shooting Times*, in an interview soon after the estate had changed hands, that his plan was to have 'more frequent shooting but to keep it sporting, and not to flood the area with huge numbers of birds'.

The locals, however, noticed the change. Pheasants were being shot just two football fields away from the churchyard: cue lurid reports of dismembered birds twitching among the graves and gunfire sounding during the services. The pursuit of death on Melangell's doorstep seemed like a particular insult to her memory, and the backlash was swift. Meetings were held, petitions signed. 'That this should be happening here is a blasphemy, an affront. Do the gunmen not see?' wrote the rock climber Jim Perrin, who had recently visited the church, in the *Country Diary* column in the Guardian.[18] Even Rowan Williams, the former Archbishop of Canterbury, intervened in the débâcle. 'The place itself is associated with an ancient legend affirming God's care for all living creatures,' he said. 'There is a profound cost to ignoring such traditions and destroying a place of genuine sanctuary.'[19]

But when I searched for more details of what had unfolded in this normally uneventful valley, I came up short. My attempts to reach out to those who had spoken out against the shooting were met with silence. Unusually, it seemed that nobody wanted to talk. So I decided to go on a pilgrimage of my own.

It was spring when I made the journey to Llangynog. The trees were not yet in leaf, although already there were primroses and stitchwort in full bloom along the verges, and lambs frolicked around the knobbled trees marking out the old field boundaries.

Evidence of the pheasants was everywhere. Feathers gleamed in the grass like burnished quills, and the gilded survivors of the previous year's carnage strutted in the fields, unaware that their ordeal was far from over. In the churchyard itself, I discovered the plastic casing of a shotgun cartridge. So much for resting in peace.

Melangell herself would have never seen a pheasant. The birds are not native to Britain: they come here only to be shot. Every year, millions of eggs and chicks are imported from Europe, whereupon they are fattened and released into the countryside in the name of field sport. Exactly how many birds are released each year onto the Llechweddygarth Estate is difficult to ascertain; the management keeps its cards close to its chest. But the figure certainly runs into the tens of thousands. Pheasants that don't succumb to shot often end up as roadkill. Driving to the church during the hunting season can feel like a particularly horrifying game of ten-pin bowling, as droves of birds scuttle before the oncoming cars.

There are plenty of reasons to dislike pheasant shooting. For those living in the vicinity of a shoot, there is the noise to consider, and the possibility of stray ammunition. Predictably, animal rights activists vehemently oppose the industrial slaughter that awaits the birds come autumn. Although Llechweddygarth does supply its birds to the local butcher – whose husband is the head keeper for the estate – many birds shot across Britain will never see the inside of a pot, destined instead for the incinerator or being fly-tipped by the roadside.[20]

Then there is the social side of it. Field sports tend to be the preserve of rich folk dressed in tweed, who are willing to part with thousands of pounds to spend a day on the moors. Those shooting at Llechweddygarth, which sits at the more expensive end of the spectrum, can expect to receive meals prepared by a Michelin-starred chef, accompanied by fine wines, champagne and cocktails. Locals speak of Londoners showing up in helicopters and heading straight to the mansion, with little evidence of financial benefit to the local community.

On top of all that, there are the environmental consequences to consider. Pheasants outweigh all of Britain's other breeding birds put together.[21] The birds are often kept in woodland pens prior to their release, where they cause long-lasting harm to the fragile environment. One study has shown that, where pheasants are present, there is more bare ground and a higher volume of nutrients in the soil as a result of their droppings. This causes the decline of ancient woodland species, including bluebells and yellow archangel, which are unable to compete with the more adaptable nettles. These flowers fail to return even after the pens have fallen into disuse.[22] Another study has observed a rise in bare ground and weed species along hedgerows, which pheasants use as corridors for crossing the countryside, with a potential knock-on effect on butterflies that depend on certain plants for food.[23]

The Llechweddygarth Estate would certainly not be alone in attracting backlash on any of these grounds. But in its conflict with the church, it was unique. Yet the more time I spent in Llangynog, hearing from locals about the saint's part in the controversy, the more complex the story became.

Everyone remembered the mounting tension that followed the intensification of the shoot – a time of heated conversations, village meetings, threats of lawyers. Media attention had only served to inflame the situation further, and there was still a sense of nervousness around the whole affair – a fear of reigniting a controversy that had been more or less put to rest by the time of my visit.

However, the sense I got, from those I spoke to, was that the uproar had never really been about the saint, but rather the usual objections to an industrial shooting estate. The people who left comments on one online petition more often objected to the disruption to the tranquillity of the valley, or the endangerment of the history and heritage embodied in the church, than the threat to the concept of sanctuary itself.[24] For those who did comment on the sacredness of the land, few seemed to come from a position of genuine belief. This is not entirely surprising: as far as I know, the residents of Llangynog are no more devout than those of any other village, while the pilgrims who visit the church, by definition, are those without personal connections to the land – people who, when the gunfire becomes too loud, can simply move on to some other saint's bones.

'It wasn't more controversial than anywhere else, but I think the people who were anti-shoot leapt onto the story of Melangell as a hook to hang their views on,' one person told me as I hung around outside the church. Another suggested that the legend had been romanticised to protest the shoot. In other words: the saint seekers had been at it again.

The anti-shoot contingent weren't the only ones at it. Around a week after the *Country Diary* column appeared in the *Guardian*, an angry refutation was printed in the letters page, written by Christopher Graffius, representing the British Association for

Shooting and Conservation. 'To pray for the aid of St Melangell for this nonsense is the real sacrilege,' he wrote. 'Better to invoke St Hubert, the patron saint of game shooting, to protect a well-run, law-abiding shoot that benefits the environment and the local area and produces good food.'[25] Graffius appeared unaware that Hubert himself had renounced the hunt after encountering a stag with a crucifix hovering between its antlers and hearing the voice of God inform him that he was headed to hell unless he changed his ways.

In fact, there is nothing in the original account of Melangell's life, composed around the fifteenth century, to suggest that the Tanat Valley was ever a refuge for animals. The *Historia Divae Monacellae* does tell of how Melangell sheltered a hare from Brochwell's dogs and that the prince created a sanctuary in her name as a result. That sanctuary, however, was originally intended for felons rather than animals – a legal practice recognised across medieval Europe. It was not wide-eyed hares but thieves and murderers that could claim protection upon this sacred land. The historian Huw Pryce suggests the re-telling of the old legend could have been a form of saint seeking in itself: an attempt by church laymen to reassert their judicial rights in the wake of the Edwardian conquest of Wales.[26]

Although, for a time, hares may have received some protection in the valley, parishioners did not extend their concern to the other creatures that shared it. Old maps show two cockpits – one in the churchyard, and another just outside the gate – where birds were forced to fight to the death for the entertainment of onlookers. The spectacles were held in May, specifically to coincide with the feast day of the saint. People congregated from miles around to join the celebrations. 'The festivals lasted, I am told, for a fortnight,' wrote the antiquarian and folklorist Reverend Elias Owen.[27]

Perhaps, in this context, the shooting of pheasants does not seem so out of place. Melangell may have been known for her love of gentle creatures, but if any of them thought they could depend on her for safe passage through the valley, they were sadly mistaken. The saint's powers had always been patchy at best; nowadays, they are weaker than ever. Without that deep sense of the sacred, the invocation of Melangell had been useless in maintaining the safety of her sanctuary. Come the shooting season, the sound of gunshot would ring out across her once-peaceful refuge, much as it always had.

Still, I was not prepared to give up on my search for hallowed ground so easily. At its core, Christianity has always been somewhat estranged from the natural world, but there have long been other gods at large in the British Isles. Lynn White Jr argued that the damage inflicted by Christianity wasn't restricted to its own doctrines of dominion, but also in pushing out the earthier religions that had been practised in older times. 'Before one cut a tree, mined a mountain, or dammed a brook, it was important to placate the spirit in charge of that particular situation, and to keep it placated,' he wrote. 'By destroying pagan animism, Christianity made it possible to exploit nature in a mood of indifference to the feelings of natural objects.'

Unwritten, ever-changing, and perched in that murky space between religion and New Age spirituality, it is hard to put a finger on these old beliefs; to separate myth from reality. But that doesn't mean that they should be dismissed. Though details may be scant, we can nonetheless be sure that people have invested the landscape with significance for thousands of years, seeking out spirits within woodlands and watercourses, marking the seasons through ritual and ceremony, telling and re-telling tales to explain the strange phenomena that heaved their world into being. The leftovers of pagan belief have always shimmered beneath the veneer of the new religion. Here, in the surroundings of Melangell's church, they burned particularly brightly.

The yew trees were the first sign of it. They were scattered among the churchyard, together forming a druidic sort of grove among the graves. The species is notoriously difficult to age, although some may have been around 2,000 years old. There is generally little evidence to support the widespread belief that yew trees were sacred to the pagans, or that their frequent appearance in churchyards was an attempt to create continuity between the old religion and the new.[28] But here it seems likely, at least, that the trees were planted to mark out a place that had already been considered sacred for a very long time. Buried beneath the present-day chapel are fragments of charcoal and cremated human bones – the remains of Bronze Age funeral pyres – hinting that this land had some ritual purpose well over a thousand years before the birth of Christ. It is not hard to imagine that Melangell might have stumbled across these graves, dotted among the yews, and felt drawn to settle by the weight of millennia of faith.

Even the church had an otherworldly feel to it. Fastened to the wall of the nave is a giant whale rib; depending on what you read, it belonged either to a giantess who lived in the mountains or to Melangell herself. Boreholes at one end suggest, only slightly less poetically, that it once formed part of a harp. Look upwards and you will find a green man peeking out from the fifteenth-century rood screen, a ribbon of oak leaves spilling from his mouth across the length of the beam. And then there are the hares: the place is mad with them all year round. There are hares sculpted into stone, hares in the gold-leaf artwork on the wall, hares woven in cloth, hares painted in watercolour for purchase by passing pilgrims. While clearly linked to the legend of the saint, it is hard to ignore the overtones of witchcraft that the creature carries in folklore.

The hills and rivers that surround the church have given rise to their own stories over the years, unconnected to the saint at its centre. The Tanat Valley has long had a reputation as a 'thin place', where the boundary between the physical and spiritual realms is easily crossed. Numerous magical beings were once believed to haunt the valley's wilder corners. A local archaeologist wrote, in 1938, of the fairies with which he was 'intimately acquainted' that emerged on the southern slopes of one mountain when twilight gathered or mists descended. One group of men, exploring caves by candlelight in 1860, claimed to have discovered a hag washing her clothes in a brass pan.[29] Legend has it, meanwhile, that the boulders at the foot of the waterfall were hurled there by giants, and that a standing stone was erected to stop a dragon causing devastation in the countryside.[30]

For Jack Hunter, who grew up in Llangynog, there is an intimate connection between ecology and spirituality. He still lives in the valley, and kindly agreed to meet me one morning to show me some of its stranger elements. An anthropologist by trade, he teaches sacred geography and contemporary cosmologies at the University of Wales Trinity Saint David. For his doctorate, he studied contemporary trance and physical mediumship, with the fieldwork involving sitting in on a number of séances in a garden shed on the outskirts of Bristol.[31] He is a member of the Fairy Investigation Society and the Afterlife Research Centre, as well as an ordained Dudeist Priest.[32]

To be frank, I wasn't sure whether I would be able to hold my own in such company: would it be considered rude to let my scepticism slip? But Jack was less outwardly eccentric than I was expecting. He showed up in a muddy 4x4, wearing a bright orange raincoat more suited to hillwalking than communing with any spirits. As we walked, he patiently answered my questions about the long history of otherworldly encounters in the valley, including his own. The first happened in a dream when he was around twelve or thirteen years old; he recalls climbing through a quarry and finding two troll-like creatures asleep in a warm cavern. The second took place after he had taken magic mushrooms for the first time, some five years later, and involved two-dimensional creatures – 'like beautiful little goblins' – walking in procession along a chest of drawers. His mother, he told me, was not particularly impressed when he confided what he had seen. But the way he talks of these visions now, at least, invites reflection: beyond the binary question of whether such beings exist in physical form, and upon the deeper mystery of how the landscape shapes the human psyche itself.

'Sometimes there's more grounding for these older folk stories – they are grounded in particular places, and relevant to a particular area,' he told me, as we circled the church. He compared the Christian story to monoculture farming: 'You bring in one system and you plant it on the land. Whereas with animist religion, it's a more polyculture thing – the stories are really rooted in particular places, and they don't necessarily translate from one location to another.'

The more I think about it, the more it makes sense. When it comes to the boulder-strewn landscape of the Tanat Valley, is a giant any less plausible a creator than a glacier, to a person who has never heard of glaciers? What were the glaciers that deposited the boulders across this valley, in any case, if not giants of sorts? Beyond that, why wouldn't you assume the strange flickers of candlelight across a cave wall was a hag on laundry day, or that the shadows in the mist were a troupe of fairies? It is easy to scoff; to disbelieve. But these are the stories that were written by the landscape itself; the happenings and inhabitants suggested by scenery seen by half-light, obscured by fog, animated by candle, set to the sound of wind and water. For the open minds who wandered such rich and wild places, they were as plausible an interpretation as any.

Even today, we are not so different from those among whom such folklore originated. The most sceptical among us can still succumb to a suggestive environment. A scary movie, watched on silent, somehow fails to frighten us. Is the fear that such music provokes really any different to the thoughts of goblins planted by a distant creaking in a wood? How many of us have felt a prickle down the spine before walking through a graveyard, despite definitely not believing in ghosts? Is the urge to hasten our footsteps across a darkening moor any different to the instinct to lower one's voice in a church, even though you have never believed in God? The landscape has a way of making you believe in its hidden layers. Even if just for a moment. Even when there's no rational cause to believe.

Jack and I headed out of the graveyard and down the road. Around a quarter of a mile from the church, where the forested hill starts to rise up from the field, is Melangell's bed – a cleft in the rock where the princess is said to have slept during her years in the wild. This is more folklore than anything else, with the first reference to it appearing during the eighteenth century. Sometimes it is known as the bed of the giantess. We ambled in its vague direction and started to scramble up the loose clods of earth among the conifers, in hope of seeking out the resting place – whomever it belonged to.

'I think a lot of stories have their origins in experiences,' said Jack, as we inched along the treeline. 'Where an ecosystem is allowed to flourish and do its own thing, that makes a place thin. Take that away, and it becomes barren. In a wilder world, there were more complex places and people possibly did have more types of experiences. And there was a place within the culture for that – even if it was just seeing a fairy.'

Reading between the lines of the *Historia Divae Monacellae*, you get a sense of how different the valley would have looked in the sixth century, back when Brochwel encountered Melangell crouching among the thorns: the first question he asked the princess was how long she had lived 'in such a wilderness'. The oak leaves carved into the medieval beams of the church indicate the richer foliage that would have been the backdrop to the encounter. Seeds and charcoal recovered from the Bronze Age cremation pits confirm that, thousands of years ago, the valley would have been partly covered by oak trees, with alders clustered along the marshy floodplain, alongside areas of open grassland and scrub.

Today, the patches of woodland and wide-girthed sycamores scattered throughout the gullies and fields provide a small sense of how the place must have looked in wilder times. Mostly, however, they reinforce how much the valley has changed. The forests and grasslands that cloak the landscape have been created and controlled by humans. Sheep patrol the fields, preventing the growth of saplings, while the sharp lines of timber plantations disrupt the flow of the hills. The mining industry has left its legacy, too, in the heavy metal contamination that lingers in the soil and river sediments and the spoil heaps on the hillsides.[33]

Perhaps it is no coincidence, then, that the fairies have also vanished from the valley. Writing in 1934, the archaeologist Robert Richards mourned that such tales were heard less frequently than they had been in the previous generation.[34] He blamed the schools – 'Can it be that our system of education has finally banished from our rural hamlets not only the terrifying ghosts which occasionally, it must be admitted, assumed most grotesque forms, but the gossamer-like fairy as well?' – but surely the changing landscape must have also played its part. The valley is still quiet, still beautiful even, but you cannot help but feel that its old spirits may have fled as industry moved in.

Beliefs wax and wane; religions rise and fall. The trouble is not that the old stories have faded. It is that no new magic has emerged to take their place. I am not sure whether cause and effect can ever be disentangled: whether it was growing disbelief – the byproduct of education, urbanisation, industrialisation and so on – that opened the door to the desecration of nature; or whether the desecration of nature came first, leading to the loss of the spirits that drew their lifeblood from its majesty and mystery. Humans have never been completely averse to hacking at the habitats of the spirits, after all. Reverence for the old gods could not save the wildwood, which was destroyed at the hands of people far more attuned to the animism of the living world than we are today. The gods of later and more destructive ages never really stood a chance.

Jack and I never managed to find Melangell's bed. But we did find something else. A few metres from where the saint was supposed to have slept was a jumble of wire fencing: the remains of an old pheasant pen, torn down and left to gather moss, here in the heart of this sanctuary, this giant's lair, this thin place. Thin, perhaps, no longer.

If we are ever to recover our sense of the sacred, we need more wild places: landscapes that set our minds whirring and our senses tingling; that leave us wondering what we saw in the half-light or heard in the wind. But we also need to reconnect with such places personally. To get to know their hidden layers.

The psychoanalyst Carl Jung was particularly concerned about the impact of the loss of myth on the human psyche. 'Thunder is no longer the voice of an angry god, nor is lightning his avenging missile. No river contains a spirit, no tree is the life principle of a man, no snake the embodiment of wisdom, no mountain cave the home of a great demon. No voices now speak to man from stones, plants, and animals, nor does he speak to them believing they can hear,' he wrote in *Man and his Symbols*, just before his death in 1961. 'His contact with nature has gone, and with it has gone the profound emotional energy that this symbolic connection supplied.'

His contact with nature has gone. The sentiment seemed to strike at the heart of the matter. Somewhere down the line, we became detached from the living world. In felling the forests, we evicted the gods that dwelled in the trees. In allowing our streams to fill with sewage, we supplanted healing with sickness. In killing so many animals, we banished the basis of future fables. The erosion of the wild from our daily lives means that the potential for otherworldly experiences has diminished. Is it still possible to regain something of that old magic? I think so: but how we go about it will always be a deeply personal mission.

The natural world is already infused with stories. Walk any path, through any mountain, along any stream, and you will find yourself in the company of old gods, shapeshifting saints, perhaps the occasional wodwo. Reviving these old tales is a good first step towards recovering that lost connection. It is also the first line of defence against those who would invoke a static narrative as a means of maintaining the status quo. The Lake District, for instance, was recently made a UNESCO World Heritage Site. To keep this status, the park authority must maintain the existing character of the landscape, which was built around the farming cultures of the past 400 years. The beauty of the open fells, however, is marred by the ecological damage wrought by sheep. If we are to preserve this landscape for its history and traditions, could we not also make room for the Neolithic axe-makers, who would have passed through wildwood as they sought out precious greenstone among the peaks of Great Langdale? For the Wild Boar of Westmorland, which supposedly had a den on Scout Scar and terrorised pilgrims during the reign of King John? For the wolves immortalised in place-names? Farming is far from the only story inscribed upon this corner of the north.

The trouble with depending on ancient myths for modern meaning, however, is that many of them leave me cold. Personally, attempts to separate authentic belief from Victorian invention or neo-pagan reinterpretation tend to inspire confusion more than connection. Stories of hags and fairies may provide some colour to a walk in the countryside, but I do not actually believe in them – and I am sure I am not unusual in that regard. Unaccompanied by genuine feeling, campaigns that attempt to leverage such tales tend to fail, either flickering out over time or leaving activists open to accusations of opportunism. What matters is not that we force ourselves to believe in outdated ideas, or carry out rituals whose allusions have been lost to time, but that we find meaning in the landscape as we encounter it today, as humans of the twenty-first century.

Like the landscape itself, stories are dynamic and ever-changing. There has never been a climax narrative: we might pretend the significance that we ourselves bestow on the land is the ultimate truth, when, in fact, the stories it holds have always been as manifold as the vegetation it has supported. Britain has produced a diverse crop of shamans, stone circles, druids, green men, goddesses, fairies, hags, saints, demons and sprites – all the products of the ecosystems, the societies and the minds in which they evolved. What we need now – what nature needs – is a new myth.

How might that myth look? For some, maybe, old stories hold as much weight as they ever did. Britain still has a thriving community of Druids, for instance, and the church has embraced the cause of environmental stewardship with growing zeal in recent years. For others, it might look like the revival and reinvention of old customs, such as wassailing or fire festivals, which act as a thread to a more magical past: a way of hitchhiking on the beliefs that flourished in wilder times. For most of us, though, there is no going back to a time, per Jung, when each river contained a spirit and voices spoke from stones. What we can do instead is rebuild our contact with nature as we encouter it today, and in doing so discover our own symbolic connections with the natural world, from which the 'profound emotional energy' of which he spoke may flow again.

It is this profound emotional energy that we see when thousands of people come together to protect a solitary tree from the chop, as happened a few years back with the Cubbington Pear, which was stationed along the route of the proposed HS2 railway line in Warwickshire. What could motivate so many people but a sense that such trees mean more than the sum of their parts? Equally, I have heard swimmers speak of the deep sense of calm that descends when they immerse themselves into an icy lake: a state of transcendence more akin to prayer or meditation than sport or adventure. The surge in enthusiasm for Britain's lost rainforests seems to hinge on something greater than a love of moss and lichen; rather, it draws upon our hope that there is still whimsy and wildness out there, if we know where to look. Without necessarily appealing to anything otherworldly, these movements treat the earth as though it were consecrated; something more than mud and leaves and atoms. It is no coincidence, I think, that they have spearheaded some of the most powerful campaigns for environmental protection in recent years.

There is one place that feels sacred to me personally, and that is Kingley Vale. The nature reserve, in the South Downs National Park, is famous for its ancient yews. Exactly how ancient, no one really knows; one theory is that they were planted to commemorate a victory over the Vikings in 850 CE. Some say that the conquered warriors still haunt the forest, and I could well believe it. The place is gloomy and ethereal all at once, the natural darkness that pools beneath the canopy enhanced by the ghostly whiteness of the chalk hills that surround it. At the top of those hills are the barrows where the Viking leaders are said to be buried.

Writing in 1906, the naturalist W. H. Hudson compared the sensation of walking among the yews to 'being in a vast cathedral; not like that of Chichester, but older and infinitely vaster, fuller of light and gloom and mystery, and more wonderful in its associations'. But the pagan vibes outweigh the Christian ones – at least, they did when my husband and I made our first trip there a couple of years ago. It was the winter solstice. We had been slow setting out for the day, beset by a sense of festive lethargy, and the sun was already low in the sky. The impression within the forest was one of premature dusk. The yew trees sprawled around us, their roots rippling through the soil, branches feeling their way through the fog.

We were not alone in having chosen this place to mark the arrival of mid-winter. There is something inherently spiritual about so many ancient yews gathered in one place. On this day, the turning point in the year, when darkness turns to light, I could sense it more acutely than ever. I felt in the midst of something, although I couldn't say what. I wanted to ask the people we passed why they had chosen to come here on this day in particular, whether they also felt this thing in their bones, if they could help me attach a name to the feeling. Beneath the boughs of one tree, I saw a woman set fire to a folded piece of paper and leave it to burn out on the wet needles, in some kind of personal ceremony. I envied her clarity: she seemed to know how to commune with the spirits in a way I did not.

For we had come here in hope of a blessing, too. My husband and I wanted a baby. We had hoped to announce one to our families this Christmas, but so far no luck. We were tired of waiting, and willing to believe in magic: to put our faith in the trees. We hoped that a small prayer, in this thin place, on the most mystical day of the year, might act as the turning point in our own lives, too.

Nine months later, on the autumn equinox – almost to the minute – our daughter was born.

EPILOGUE

Old Souls

She enters the world beneath the bright lights of an operating theatre, pulled out of my uterus by a doctor whose face I cannot remember. I do not witness the moment she opens her eyes for the first time: I am lying behind a screen, high on a cocktail of painkillers, mumbling incomprehensibly about pigeons and netball.

Her birth is quick and clinical – not how I imagined it would be – yet the magic in that moment is undeniable. I have spent the months of my pregnancy, and many years before it, researching and writing this book. For so long, my mind has been absorbed by thoughts of megafauna, wildwood, bison, eagles, yew and pine; by sadness over how many wonders have been lost from this planet since humans gained the upper hand. The fact that my baby and I have emerged alive and healthy from my long and difficult labour is a reminder of how much we have also gained.

Later that day, once the drugs have worn off, I muse on what will become of this tiny human. I realise there is a good chance she will see the end of the century, and wonder what the world will look like by then. Will there still be corncrakes in the meadows and sharks in the sea? Will the midsummer sun be too much to bear? Will it still be possible to know the enchantments of an ancient woodland, a wild orchid, the winter snow? My despair at thousands of years of destruction condenses into a kernel of fear for this one little girl. I am used to dealing with scientific projections about years beyond my lifespan. The fact that I will not have to deal personally with the catastrophe to come has always helped to mute the pain. Now, for the first time, the future no longer feels abstract. It is terrifyingly real.

But it is not only the future that feels tangible. The past draws closer, too. The experience of pregnancy and childbirth has added me to a lineage stretching back hundreds of thousands of years. My growing belly, the feet in my ribs, the agony of contractions – each is a reminder that time is an artificial construct. The landscape around us may have changed, but the experience of being a human has not. The process of growing and birthing my daughter was the same, in essence, to that which millions of women have undergone before me. Only the setting had changed. Had I been alive at almost any other time in human history, my daughter would have emerged straight from the womb and into the wilderness. She would have arrived on this earth to the sound of crackling firewood or morning birdsong, the smell of blood and placenta dangerously attractive to any passing predators.

Throughout this book, I have advocated for the return of an older world for two reasons. One, because the landscapes of the past were richer and more abundant than those of today, and I believe we have a moral duty to re-create those qualities for the sake of nature itself. Humans have taken too much. It is time to give back some of what we have destroyed. In my view, there is no better blueprint for that than history. Conservationists often point out that we cannot return the earth to a lost golden age, and I agree with that. But that should not equate to a refusal to look backwards at all: the future can be both ancient and modern. In fossils, pollen and poetry we might find a guide to the task. In places like Carrifran, the Gower Peninsula or the dunescape of Kraansvlak, we can already see that it is possible. Old records show us how the landscape looked when its ecosystems were still intact; where plants grew and how animals behaved before humans pulled all the strings. More than that, however, they show us the unparalleled beauty of a world that has long been lost from living memory. Perhaps the greatest danger of ignoring the past is that we forget how magnificent the earth can be, and accept too little from the future as a result.

The second reason is that, by looking to the past, we can reinvigorate our own relationship with the natural world. Throughout this book, I have showcased just a few examples of the entanglements that still exist between people and nature: places where humans can still claim to be something like the keystone species we were throughout much of history. For most of us, however, such closeness no longer exists. Mending this fracture does not mean we must all revert to hunter-gathering or even small-scale farming for a living – an unrealistic and undesirable goal. But it does require us to take a more ancient perspective of the earth: not as a distinct and distant entity, but as something woven through our being. A source of food and also wonder. Of money and also magic. Of materials and also identity. It means reconceiving our place in the natural order, so that we are not just admirers or even protectors of nature, but also participants in its cycles and processes: as the mammals we really are, or as proxies for the megafauna we wiped out; as managers, as foragers and fishers, or even as storytellers who might translate the old narratives of the land into the language of the modern day. There remains a place for us in the wild, if we can only rewild ourselves.

During the four nights I spent in hospital, however, a third reason occurred to me. In the midst of those blurry, tired, milky days, what seemed most important of all was that my new daughter should have the chance to live a happy and meaningful life. That she might experience the full spectrum of emotions that the world has to offer. She could have a hundred years on this earth, and I don't want any of them to be diminished by the mistakes that her predecessors have made.

For a short moment in time, my baby is an empty vessel, ready to receive whatever the world pours into her. She has it in her to become both mammoth hunter and rocket scientist. It may be the former, however, that comes more naturally – even as society sets her down a different course. Evolutionary psychologists argue that, for all the advances of the past 10,000 years, the human brain still belongs in the ancient world. The primeval daughter of my imagination would have had forests and grasslands for her playground, where she would have become attuned to the stimuli of the wild: her ears primed for the footfall of predators, her eyes quick to spot threats on the horizon, her skin sensitive to a coming storm. Over time, that playground would have been transformed into pharmacy, toolshed and larder, where she might have learned to make poultices from leaves, brew tea from herbs, harvest reeds for bedding, and much more besides. The better she absorbed such lessons, the greater her chance of survival.

Evolutionary psychology holds that our ancestors passed down the skills needed to navigate the demands of the wilderness until they eventually became encoded in our genes. After this, humans no longer needed to learn these behaviours. They became a set of instincts bestowed upon us from birth, unveiling themselves as we passed through life and encountered new challenges and threats. It is the psychological extension to Darwin's theory of evolution: an acknowledgment that our minds, as well as our bodies, have been shaped through millennia of natural selection. Or, as Leda Cosmides and John Tooby, two of the foremost proponents of evolutionary psychology put it: our modern skulls house a Stone Age mind.[1]

In 1984, the American biologist E. O. Wilson coined the term biophilia to describe 'the innately emotional affiliation of human beings to other living organisms'. Later, in a volume co-edited by Wilson, the social ecologist Stephen Kellert expanded on the evolutionary underpinnings for this hypothesis, showing how connecting with nature would have stacked the odds in our favour. Most obvious was that intimacy with the environment gave humans access to a 'vast cornucopia of food, medicines, clothing, tools, and other material benefits', he wrote. These provided sustenance, protection and security, enhancing the likelihood that people who nurtured such an intimacy would pass their genes onto the next generation. Those who felt awe and fascination with the natural world, meanwhile, were more likely to acquire the knowledge that bestowed such benefits.[2]

The biophilia hypothesis is often characterised as the evolutionary underpinning for our positive responses to nature: why we feel calm and restored by a natural setting. The utilitarian approach of our prehistoric ancestors has translated into what today we would more commonly call love. In modern society, this manifests as holidays in the mountains, weekends in city parks, flowers on the kitchen table. The 'nature cure' has been embraced to such an extent that NHS doctors now prescribe outdoor activities, including community gardening and open-water swimming, to improve both mental and physical health.

Wilson was clear, however, that biophilia was not only about rest and relaxation, but encompassed a range of emotions 'from attraction to aversion, from awe to indifference, from peacefulness to fear-driven anxiety'.[3] It is easy to understand why the focus has been on the positive end of the spectrum. Nature is already in trouble. Why emphasise past conflicts when there are so many in the present? It is convenient to forget that bears were feared as well as revered; that plants could poison as well as nourish; that forests invoked dread as well as wonder. Yet it seems that these experiences, too, were written into our genes.

In his 1897 paper, *A Study of Fears*, the psychologist Stanley Hall concluded that the terrors of contemporary children could act as a guide to the inner lives of our ancestors.[4] Through a questionnaire sent to parents and teachers, he gathered the childhood fears of more than 1,700 people. He then analysed this list for recurrent themes. The most common by far was fear of animals. The examples are striking. It is not hard to imagine them as the internal monologue of a hunter wandering the wilderness in search of a meal.

For months had bears on the brain, fancying them in the next room.
Had for years fear of being carried off by an eagle.
Imagines wolves' eyes in all dark corners.
The horror of cats is that they are sly, noiseless, witch-like, shiny-eyed, and you never know what they will do next.
Shudders at every rustling sound in the woods made by the wind in trees, thinking it a rattlesnake.

Following animals was a fear of celestial phenomena, with thunder and lightning considered most terrifying of all, while darkness was thought scarier than death. Such fears, Hall noted, would have made sense among the primeval children who grew up immersed in the natural world, but were out of proportion to the dangers that a nineteenth-century child was actually likely to face. 'Night is now the safest time, serpents are no longer among our most fatal foes, and most of the animal fears do not fit the present conditions of civilised life,' he wrote. 'The weather fears and the incessant talk about weather fits a condition of life in trees, caves or tents, or at least of far greater exposure, and less protection from heat, cold, storm, etc., than present houses, carriages and even dress afford.'

Although conducted more than a century ago, Hall's research has held up surprisingly well.[5]

More recent research has confirmed that humans are quick to detect the dangers of the distant past, even when they are irrelevant to us today.[6] Numerous studies have confirmed the intensity of animal-related fear among young children. A 2012 survey, conducted by the ChildFund Alliance, found that fears of insects and dangerous animals outstripped fears of death, disease, war and the end of the world. This remained consistent among children from both developed and developing countries, suggesting roots in something deeper than the chance of actually meeting a tiger, stepping on a scorpion or contracting malaria.[7]

Fear, however, does not remain static throughout our lives. Further research has shown that our instinctive terrors, designed to keep us safe in the ancient world, are ultimately joined by a set of learned ones, designed to protect us from the more realistic dangers of modern times. As we grow up, the horrors of animals, thunder and darkness subside, only to be replaced by those of our immediate environment. School. Car accidents. Violence. Syringes. Guns.[8]

I would rather my daughter had nothing to fear at all, but I know I cannot protect her forever. And so, more than that, I want to preserve some part of her old soul. No doubt some of it has withered already – as it must. As we head into winter, more of our time is spent indoors. We are sheltered from the rain and unbuffeted by wind. Thunder sounds distant, even when it is close. Electricity blurs the boundary between day and night. The most ferocious animal that the baby encounters is our household cat. I know that already her mind is being shaped by the world we have created for her, and I worry what this will mean for her psyche. Will she ever feel at home in a world where her primeval instincts must be denied at every turn? Can any of us?

My daughter will never rely on the wild as her ancestors once did. She will never know how it feels to walk through the wildwood or come face-to-face with a mammoth. Her world, in a sense, will always be dimmed. But perhaps it is still possible to resurrect a few of nature's ghosts. To restore something of that old abundance, and with it to salvage something of the emotional intensity that surely marked the majority of our years on this planet. I believe it is. I have seen it. I have felt it.

I want my daughter to know pinewoods, meadows, winding rivers and frozen lakes. I want her to experience the wonder of an eagle upon a crag. To delight in a beaver building a dam. To feel the electricity of a kingfisher darting downstream. And fear, if she must feel it – let it not be of guns and cars, but of howls in the distance; of shadows in the forest; of eyes in the dark. Let her heart race. Let her feel alive.

Notes

ONE: *When Humans Were Wild*

1. J. A. J. Gowlett, 'The Discovery of Fire by Humans: A Long and Convoluted Process', *Philosophical Transactions of the Royal Society B: Biological Sciences* 371, no. 1696 (5 June 2016): 20150164, https://doi.org/10.1098/rstb.2015.0164.
2. Nicholas R. Longrich, 'When Did We Become Fully Human? What Fossils and DNA Tell Us About the Evolution of Modern Intelligence', *The Conversation* (9 September 2020) http://theconversation.com/when-did-we-become-fully-human-what-fossils-and-dna-tell-us-about-the-evolution-of-modern-intelligence-143717.
3. John Aubrey, 'An Essay towards the Description of the North Division of Wiltshire.' *Wiltshire: The Topographical Collections of John Aubrey* (1838).
4. Erle C. Ellis et al., 'People Have Shaped Most of Terrestrial Nature for at Least 12,000 Years', *Proceedings of the National Academy of Sciences* 118, no. 17 (27 April 2021): e2023483118, https://doi.org/10.1073/pnas.2023483118.
5. Some people argue that the 'Holocene' is actually just another interglacial period within the ongoing Pleistocene.
6. James B. Innes and Jeffrey J. Blackford, 'Disturbance and Succession in Early to Mid-Holocene Northern English Forests:

Palaeoecological Evidence for Disturbance of Woodland Ecosystems by Mesolithic Hunter-Gatherers', *Forests* 14, no. 4 (April 2023): 719, https://doi.org/10.3390/f14040719.

7. J. B. Innes and J. J. Blackford, 'The Ecology of Late Mesolithic Woodland Disturbances: Model Testing with Fungal Spore Assemblage Data', *Journal of Archaeological Science* 30, no. 2 (1 February 2003): 185–94, https://doi.org/10.1006/jasc.2002.0832.
8. Paul Mellars, 'Fire Ecology, Animal Populations and Man: A Study of Some Ecological Relationships in Prehistory', *Proceedings of the Prehistoric Society* 42 (December 1976): 15–45, https://doi.org/10.1017/S0079497X00010689.
9. Daniel Groß et al., 'Adaptations and Transformations of Hunter-Gatherers in Forest Environments: New Archaeological and Anthropological Insights', *The Holocene* 29, no. 10 (1 October 2019): 1531–44, https://doi.org/10.1177/0959683619857231.
10. Peter Rowley-Conwy and Robert Layton, 'Foraging and Farming as Niche Construction: Stable and Unstable Adaptations', *Philosophical Transactions of the Royal Society B: Biological Sciences* 366, no. 1566 (27 March 2011): 849–62, https://doi.org/10.1098/rstb.2010.0307.
11. W. Ian Montgomery et al., 'Origin of British and Irish Mammals: Disparate Post-Glacial Colonisation and Species Introductions', *Quaternary Science Reviews* 98 (15 August 2014): 144–65, https://doi.org/10.1016/j.quascirev.2014.05.026.
12. Graeme Warren et al., 'The Potential Role of Humans in Structuring the Wooded Landscapes of Mesolithic Ireland: A Review of Data and Discussion of Approaches', *Vegetation History and Archaeobotany* 23, no. 5 (1 September 2014): 629–46, https://doi.org/10.1007/s00334-013-0417-z.
13. Warren et al., 'The potential role of humans in structuring the wooded landscapes of Mesolithic Ireland'.
14. Paul Davies, John G. Robb and Dave Ladbrook, 'Woodland Clearance in the Mesolithic: The Social Aspects', *Antiquity* 79, no. 304 (June 2005): 280–88, https://doi.org/10.1017/S0003598X00114085.

15. Elisabeth S. Bakker et al., 'Combining Paleo-Data and Modern Exclosure Experiments to Assess the Impact of Megafauna Extinctions on Woody Vegetation', *Proceedings of the National Academy of Sciences of the United States of America* 113, no. 4 (26 January 2016): 847–55, https://doi.org/10.1073/pnas.1502545112.
16. Norman Owen-Smith, 'Pleistocene Extinctions: The Pivotal Role of Megaherbivores', *Paleobiology* 13, no. 3 (1987): 351–62.
17. Christopher Sandom et al., 'Global Late Quaternary Megafauna Extinctions Linked to Humans, Not Climate Change', *Proceedings of the Royal Society B: Biological Sciences* 281, no. 1787 (22 July 2014): 20133254, https://doi.org/10.1098/rspb.2013.3254.
18. Tim Flannery, *Europe: The First 100 Million Years* (Penguin Books, 2019): 219.
19. Michael R. Waters et al., 'Pre-Clovis Mastodon Hunting 13,800 Years Ago at the Manis Site, Washington', *Science* 334, no. 6054 (21 October 2011): 351–53, https://doi.org/10.1126/science.1207663; Pavel Nikolskiy and Vladimir Pitulko, 'Evidence from the Yana Palaeolithic Site, Arctic Siberia, Yields Clues to the Riddle of Mammoth Hunting', *Journal of Archaeological Science* 40, no. 12 (1 December 2013): 4189–97, https://doi.org/10.1016/j.jas.2013.05.020; Karina V. Chichkoyan et al., 'Description and Interpretation of a *Megatherium Americanum* Atlas with Evidence of Human Intervention', *Rivista Italiana Di Paleontologia e Stratigrafia* 123, no. 1 (2017), https://doi.org/10.13130/2039-4942/8019.
20. S. A. Zimov et al., 'Steppe-Tundra Transition: A Herbivore-Driven Biome Shift at the End of the Pleistocene', *The American Naturalist* 146, no. 5 (1995): 765–94.
21. Susan Rule et al., 'The Aftermath of Megafaunal Extinction: Ecosystem Transformation in Pleistocene Australia', *Science* 335, no. 6075 (23 March 2012): 1483–86, https://doi.org/10.1126/science.1214261.
22. Jacquelyn L. Gill et al., 'Pleistocene Megafaunal Collapse, Novel Plant Communities, and Enhanced Fire Regimes in North

America', *Science* 326, no. 5956 (20 November 2009): 1100–1103, https://doi.org/10.1126/science.1179504.

23. Owen-Smith, 'Pleistocene Extinctions'.
24. Mauro Galetti et al., 'Ecological and Evolutionary Legacy of Megafauna Extinctions', *Biological Reviews* 93, no. 2 (2018): 845–62, https://doi.org/10.1111/brv.12374.
25. Galetti et al., 'Ecological and Evolutionary Legacy of Megafauna Extinctions'.
26. Galetti et al., 'Ecological and Evolutionary Legacy of Megafauna Extinctions'.
27. Daniel H. Janzen and Paul S. Martin, 'Neotropical Anachronisms: The Fruits the Gomphotheres Ate', *Science* 215, no. 4528 (1982): 19–27.
28. Logan Kistler et al., 'Gourds and Squashes (Cucurbita Spp.) Adapted to Megafaunal Extinction and Ecological Anachronism through Domestication', *Proceedings of the National Academy of Sciences* 112, no. 49 (8 December 2015): 15107–12, https://doi.org/10.1073/pnas.1516109112.
29. Jessica C. Thompson et al., 'Early Human Impacts and Ecosystem Reorganization in Southern-Central Africa', *Science Advances* 7, no. 19 (5 May 2021): eabf9776, https://doi.org/10.1126/sciadv.abf9776.
30. Wil Roebroeks et al., 'Landscape Modification by Last Interglacial Neanderthals', *Science Advances* 7, no. 51 (15 December 2021): eabj5567, https://doi.org/10.1126/sciadv.abj5567.
31. Eduard Pop and Corrie Bakels, 'Semi-Open Environmental Conditions during Phases of Hominin Occupation at the Eemian Interglacial Basin Site Neumark-Nord 2 and Its Wider Environment', *Quaternary Science Reviews* 117 (1 June 2015): 72–81, https://doi.org/10.1016/j.quascirev.2015.03.020.

TWO: *Small and Beautiful Clues*

1. Henry Montague Grover, *A Voice from Stonehenge* (W.J. Cleaver, 1847): 5.
2. Sharon Turner, *The History of the Anglo-Saxons* (1823): 8.

3. Alfred A. Walton, *History of the Landed Tenures of Great Britain & Ireland, from the Norman Conquest to the Present Time: Dedicated to the People of the United Kingdom* (Charles H. Clarke, 13, Paternoster Row, 1865): 28.
4. W. G. Hoskins, *The Making of the English Landscape* (Little Toller Books, 2013).
5. H. J. Fleure, 'The Racial History of the British People', *Geographical Review* 5, no. 3 (1918): 216–31, https://doi.org/10.2307/207642.
6. The other notable early proponent of this theory was O.G.S. Crawford, a pioneer of aerial archaeology, who wrote as early as 1922 that 'primitive man selected those regions which were free from dense forest and marsh and which also provided good open pasture land to graze his flocks and herds upon'. The renowned botanist Arthur Tansley similarly suggested that the chalk downlands and sandy soils would have been free from forest in his 1911 work, *Types of British Vegetation*.
7. H.J. Fleure and Wallace Whitehouse, 'Early Distribution and Valleyward Movement of Population in South Britain', *Archaeologica Cambrensis* 6 vol. XVI (April 1916): 133.
8. Sir Cyril Fox, *The Personality of Britain: Its Influence on Inhabitant and Invader in Prehistoric and Early Historic Times* (National Museum of Wales, 1959): 64.
9. Fox, *The Personality of Britain*: 55.
10. It is often attributed to Strabo, but this appears to have no basis in fact. The earliest reference I have found to the best-known phrasing – that Britain once 'had so many trees that a squirrel could go from John O'Groats to Land's End without touching the ground' – comes from the 2013 book, *1,339 QI Facts To Make Your Jaw Drop*.
11. Maurice B. Adams, 'Suggestions Towards an Appreciation of the Picturesque Considered in Relation to Social Conditions and Environment', *Journal of the Royal Society of Arts* 66, no. 3408 (1918): 292–304.
12. W. F. Grimes, 'The Megalithic Monuments of Wales', *Proceedings of the Prehistoric Society* 2, no. 1–2 (January 1936): 106–39, https://doi.org/10.1017/S0079497X00021691.

13. Fox, *The Personality of Britain*: 58.
14. Alice Garnett, 'The Loess Regions of Central Europe in Prehistoric Times', *The Geographical Journal* 106, no. 3/4 (1945): 132–43, https://doi.org/10.2307/1789265.
15. J. Troels-Smith, 'Johannes Iversen', *Bulletin of the Geological Society of Denmark*, Vol. 24/01-02 (1975): 113–125.
16. Johs. Iversen, 'Moorgeologische untersuchungen auf Grönland', *Meddelelser fra Dansk Geologisk Forening* 8, no. 4 (1934): 341–358.
17. Christen Leif Vebæk, 'Nordboforskningen i Grønland. Resultater Og Fremtidsopgaver.', *Geografisk Tidsskrift*, 1 January 1943, https://tidsskrift.dk/geografisktidsskrift/article/view/48352.
18. Johs. Iversen, 'Origin of the Flora of Western Greenland in the Light of Pollen Analysis', *Oikos* 4, no. 2 (1952): 85–103, https://doi.org/10.2307/3564805.
19. As we saw in the previous chapter, this view of early hunter-gatherers has since been somewhat revised.
20. Johs. Iversen, 'Landnam i Danmarks Stenalder. En Pollenanalytisk Undersøgelse over Det Første Landbrugs Indvirkning Paa Vegetationsudviklingen', *Danmarks Geologiske Undersøgelse II. Række* 66 (31 December 1941): 1–68, https://doi.org/10.34194/raekke2.v66.6855.
21. Harry Godwin, 'Obituary: Tribute to Four Botanists', *The New Phytologist* 72, no. 5 (1973): 1245–50.
22. N. Roberts et al., 'Europe's Lost Forests: A Pollen-Based Synthesis for the Last 11,000 Years', *Scientific Reports* 8, no. 1 (15 January 2018): 716, https://doi.org/10.1038/s41598-017-18646-7.
23. F. W. M. Vera, *Grazing Ecology and Forest History* (CABI Pub., 2000): 356–7.
24. Isabella Tree, *Wilding: The Return of Nature to a British Farm* (Pan Macmillan, 2018): 91.
25. Fraser J. G. Mitchell, 'How Open Were European Primeval Forests? Hypothesis Testing Using Palaeoecological Data', *Journal of Ecology* 93, no. 1 (2005): 168–77, https://doi.org/10.1111/j.1365-2745.2004.00964.x.

26. László Demeter et al., 'Rethinking the Natural Regeneration Failure of Pedunculate Oak: The Pathogen Mildew Hypothesis', *Biological Conservation* 253 (1 January 2021): 108928, https://doi.org/10.1016/j.biocon.2020.108928.
27. Richard H. W Bradshaw, Gina E Hannon and Adrian M Lister, 'A Long-Term Perspective on Ungulate–Vegetation Interactions', *Forest Ecology and Management*, Forest Dynamics and Ungulate Herbivory: From Leaf to Landscape, 181, no. 1 (3 August 2003): 267–80, https://doi.org/10.1016/S0378-1127(03)00138-5.
28. Nicki J. Whitehouse and David Smith, 'How Fragmented Was the British Holocene Wildwood? Perspectives on the "Vera" Grazing Debate from the Fossil Beetle Record', *Quaternary Science Reviews* 29, no. 3 (1 February 2010): 539–53, https://doi.org/10.1016/j.quascirev.2009.10.010.
29. Christopher J. Sandom et al., 'High Herbivore Density Associated with Vegetation Diversity in Interglacial Ecosystems', *Proceedings of the National Academy of Sciences* 111, no. 11 (18 March 2014): 4162–67, https://doi.org/10.1073/pnas.1311014111.
30. Elena A. Pearce et al., 'Substantial Light Woodland and Open Vegetation Characterized the Temperate Forest Biome before Homo Sapiens', *Science Advances* 9, no. 45 (10 November 2023): eadi9135, https://doi.org/10.1126/sciadv.adi9135.
31. Isabella Tree, 'If You Want to Save the World, Veganism Isn't the Answer', *The Guardian* (25 August 2018), https://www.theguardian.com/commentisfree/2018/aug/25/veganism-intensively-farmed-meat-dairy-soya-maize.
32. Vera, *Grazing Ecology and Forest History*: 381.
33. Pieter Hotse Smit, 'Frans Vera kan niet langer zwijgen over "zijn" Oostvaardersplassen: "Schande dat de politiek is gezwicht"', *de Volkskrant* (24 January 2020), https://www.volkskrant.nl/nieuws-achtergrond/frans-vera-kan-niet-langer-zwijgen-over-zijn-oostvaardersplassen-schande-dat-de-politiek-is-gezwicht~ba3ce1ef/.
34. Frans Vera [@FransVera], *Twitter* (10 November 2022), https://twitter.com/FransVera/status/1590744692790296577.

THREE: *Return of the Native*

1. Myrtle and Philip Ashmole, *The Carrifran Wildwood Story: Ecological Restoration from the Grass Roots* (Borders Forest Trust, 2009).
2. George F. Peterken, *Natural Woodland: Ecology and Conservation in Northern Temperate Regions* (Cambridge University Press, 1996): 381.
3. Peterken, *Natural Woodland*: 381.
4. Peterken, *Natural Woodland*: 370.
5. Rewilding Britain, *Reforesting Britain: Why natural regeneration should be our default approach to woodland expansion* (2020), https://www.rewildingbritain.org.uk/about-us/what-we-say/research-and-reports/reforesting-britain.
6. Joshua Bauld et al., 'Assessing the Use of Natural Colonization to Create New Forests within Temperate Agriculturally Dominated Landscapes', *Restoration Ecology* 31, no. 8 (2023): e14004, https://doi.org/10.1111/rec.14004.
7. Andrew Bachel, 'West Affric: learning from the past – new woods in Affric', *Scottish Woodland History Discussion Notes IV* (Centre for Environmental History and Policy, 1999): 22–28.
8. Joan McAlpine, Scottish Parliament Motion S5M-13223 (July 2018), https://www.parliament.scot/chamber-and-committees/votes-and-motions/votes-and-motions-search/S5M-13223.
9. Kornelis Jan Willem Oostheok, 'An Environmental History of State Forestry in Scotland, 1919–1970', 2001, http://dspace.stir.ac.uk/handle/1893/1450; T. C. Smout, *History of the Native Woodlands of Scotland 1500–1920* (Edinburgh University Press, 2007); J.H. Dickson, 'Scottish Woodlands: Their Ancient Past and Precarious Present', *Botanical Journal of Scotland* 46, no. 2 (1 January 1992): 155–65, https://doi.org/10.1080/03746600508684785.
10. 'West Highland Survey. An Essay in Human Ecology', *AIBS Bulletin* 5, no. 3 (1 July 1955): 192, https://doi.org/10.1093/aibsbulletin/5.3.7-c.
11. Peter D. Moore, 'Origin of Blanket Mires', *Nature* 256, no. 5515 (July 1975): 267–69, https://doi.org/10.1038/256267a0.

12. Scottish Green Party, *A Rural Manifesto for the Highlands* (Highland Green Party, Land-Use Working Group, 1989).
13. T.J. Sloan et al., 'Peatland Afforestation in the UK and Consequences for Carbon Storage', *Mires and Peat*, no. 23 (19 October 2018): 1–17, https://doi.org/10.19189/MaP.2017.OMB.315.
14. D. A. Stroud et al., 'Birds, Bogs and Forestry: The Peatlands of Caithness and Sutherland', Technical report (Nature Conservancy Council, 1988), http://www.jncc.gov.uk/page-4322.
15. Brian Holtam, 'Forestry Practice in Britain Is Applied Terrestrial Ecology', *The Commonwealth Forestry Review* 55, no. 2 (164) (1976): 123–27.
16. Frank Fraser Darling et al., 'The Tenth Commonwealth Forestry Conference', *The Commonwealth Forestry Review* 53, no. 4 (158) (1974): 251–79.
17. Richard Tipping, 'The Form and the Fate of Scotland's Woodlands', *Proceedings of the Society of Antiquaries of Scotland* 124 (30 November 1995): 1–54, https://doi.org/10.9750/PSAS.124.1.54.
18. Smout, *History of the Native Woodlands of Scotland 1500–1920.*
19. A. V. Gallego-Sala et al., 'Climate-Driven Expansion of Blanket Bogs in Britain during the Holocene', *Climate of the Past* 12, no. 1 (28 January 2016): 129–36, https://doi.org/10.5194/cp-12-129-2016.
20. Richard Tipping, 'Blanket Peat in the Scottish Highlands: Timing, Cause, Spread and the Myth of Environmental Determinism', *Biodiversity and Conservation* 17, no. 9 (1 August 2008): 2097–2113, https://doi.org/10.1007/s10531-007-9220-4; Natural England, '*The Historic Peat Record: Implications for the Restoration of Blanket Bog – NEER011*' (2016), https://publications.naturalengland.org.uk/publication/5155418650181632.
21. Oliver Rackham, *Trees and Woodland in the British Landscape* (Hachette UK, 2020): 25.
22. Mairi J. Stewart, 'Does the Past Matter in Scottish Woodland Restoration?', in *Restoration and History* (Routledge, 2009).
23. Bachel, 'West Affric: learning from the past – new woods in Affric'.

24. Richard Tipping, Althea Davies and Eileen Tisdall, 'Long-Term Woodland Dynamics in West Glen Affric, Northern Scotland', *Forestry: An International Journal of Forest Research* 79, no. 3 (1 July 2006): 351–59, https://doi.org/10.1093/forestry/cpl022.
25. Scottish Woodland History Discussion Group, *Scottish Woodland History Discussion Notes IV* (Centre for Environmental History and Policy, 1999).
26. Alan Watson Featherstone, 'An Oasis of Life in a Depleted Landscape', 19 September 2015, https://alanwatsonfeatherstone.com/an-oasis-of-life-in-a-depleted-landscape/.
27. Emily Warner et al., 'Does Restoring Native Forest Restore Ecosystem Functioning? Evidence from a Large-Scale Reforestation Project in the Scottish Highlands', *Restoration Ecology* 30, no. 3 (2022): e13530, https://doi.org/10.1111/rec.13530.
28. Emily Warner et al., 'The Response of Plants, Carabid Beetles and Birds to 30 Years of Native Reforestation in the Scottish Highlands', *Journal of Applied Ecology* 58, no. 10 (2021): 2185–94, https://doi.org/10.1111/1365-2664.13944.
29. Christopher Smout, 'The History and the Myth of the Scots Pine', *RSFS Scottish Forestry* vol. 68 no. 1 (2014).

FOUR: *Holocene Farm*

1. Sacha E. Davis, 'Hospitality Networks, British Travel Writers, and the Dissemination of Competing Transylvanian Claims to Civilization, 1830s–1930s', *Nationalities Papers* 46, no. 4 (July 2018): 612–32, https://doi.org/10.1080/00905992.2018.1448375.
2. Katarina Gephardt, '"The Enchanted Garden" or "The Red Flag": Eastern Europe in Late Nineteenth-Century British Travel Writing', *Journal of Narrative Theory* 35, no. 3 (2005): 292–306.
3. Charles Boner, *Transylvania: Its Products and Its People* (Longmans, Green, 1865): 75.
4. Emily Gerard, *The Land Beyond the Forest: Facts, Figures, and Fancies from Transylvania* (W. Blackwood and Sons, 1888).

5. Andrew Fox, 'Trajanic Trees: the Dacian Forest on Trajan's Column', *Papers of the British School at Rome* 87 (October 2019): 47–69, https://doi.org/10.1017/S006824621800034X.
6. WWF, 'The Ecological Effects of Mining Spills in the Tisza River System in 2000' (April 2002), Microsoft Word – Cyanide Report First ReviewFINAL2.doc (panda.org); Nick Thorpe, '*Legacy of Tisza Poisoning*', (BBC 31 January 2002), http://news.bbc.co.uk/1/hi/world/europe/1794227.stm.
7. Climate ADAPT, 'Adaptation in Carpathian Mountains – English', accessed 22 January 2024, https://climate-adapt.eea.europa.eu/en/countries-regions/transnational-regions/carpathian-mountains/impacts.
8. Paul Dragos Aligica and Adina Dabu, 'Land Reform and Agricultural Reform Policies in Romania's Transition to the Market Economy: Overview and Assessment', *Eastern European Economics* 41, no. 5 (2003): 49–69.
9. Ole Henrik Magga, 'Diversity in Saami Terminology for Reindeer, Snow, and Ice', *International Social Science Journal* 58, no. 187 (2006): 25–34, https://doi.org/10.1111/j.1468-2451.2006.00594.x.
10. Cosmin Ivaşcu and László Rákosy. '2. Biocultural adaptations and traditional ecological knowledge in a historical village from Maramureş land, Romania', *Knowing our Lands and Resources* (UNESCO 2017): 20.
11. Ioana-Ruxandra Fruntelată, Cristian Muşa and Elena Dudău, 'Haylife and Haylore in Starchiojd (Prahova county, Romania): from Present to Past', *Martor* 21 (2016): 87–100; Anamaria Iuga. 'Intangible Hay Heritage in Şurdeşti', *Martor* 21 (2016): 67–84; Cosmin Marius Ivaşcu, Kinga Öllerer and László Rákosy, 'The Traditional Perceptions of Hay and Hay-Meadow Management in a Historical Village from Maramureş County, Romania', *Martor* 21 (2016): 39–51.
12. Marlene Roellig et al., 'Brown Bear Activity in Traditional Wood-Pastures in Southern Transylvania, Romania', *Ursus* 25, no. 1 (May 2014): 43–52, https://doi.org/10.2192/URSUS-D-13-00007.1.

13. Ine Dorresteijn et al., 'The Conservation Value of Traditional Rural Landscapes: The Case of Woodpeckers in Transylvania, Romania', *PLOS ONE* 8, no. 6 (19 June 2013): e65236, https://doi.org/10.1371/journal.pone.0065236; Tibor Hartel et al., 'Abundance of Large Old Trees in Wood-Pastures of Transylvania (Romania)', *Science of The Total Environment* 613–614 (1 February 2018): 263–70, https://doi.org/10.1016/j.scitotenv.2017.09.048.
14. Jacqueline Loos et al., 'Low-Intensity Agricultural Landscapes in Transylvania Support High Butterfly Diversity: Implications for Conservation', *PLOS ONE* 9, no. 7 (24 July 2014): e103256, https://doi.org/10.1371/journal.pone.0103256.
15. 'Natural History – Fundatia Adept', 11 March 2018, https://fundatia-adept.org/natural-history/.
16. Angelica Feurdean et al., 'Origin of the Forest Steppe and Exceptional Grassland Diversity in Transylvania (Central-Eastern Europe)', *Journal of Biogeography* 42, no. 5 (2015): 951–63, https://doi.org/10.1111/jbi.12468; Angelica Feurdean et al., 'Biodiversity-Rich European Grasslands: Ancient, Forgotten Ecosystems', *Biological Conservation* 228 (1 December 2018): 224–32, https://doi.org/10.1016/j.biocon.2018.09.022.
17. Wolfgang Willner et al., 'Long-Term Continuity of Steppe Grasslands in Eastern Central Europe: Evidence from Species Distribution Patterns and Chloroplast Haplotypes', *Journal of Biogeography* 48, no. 12 (2021): 3104–17, https://doi.org/10.1111/jbi.14269.
18. Feurdean et al., 'Origin of the forest steppe and exceptional grassland diversity in Transylvania (central-eastern Europe)'.
19. M. Hejcman et al., 'Origin and History of Grasslands in Central Europe – a Review', *Grass and Forage Science* 68, no. 3 (2013): 345–63, https://doi.org/10.1111/gfs.12066.
20. Angelica Feurdean et al., 'Biodiversity Variability across Elevations in the Carpathians: Parallel Change with Landscape Openness and Land Use', *The Holocene* 23, no. 6 (1 June 2013): 869–81, https://doi.org/10.1177/0959683612474482.
21. Jessie Woodbridge et al., 'What Drives Biodiversity Patterns? Using Long-Term Multidisciplinary Data to Discern Centennial-

Scale Change', *Journal of Ecology* 109, no. 3 (2021): 1396–1410, https://doi.org/10.1111/1365-2745.13565; Roy van Beek, Dominique Marguerie and Francoise Burel, 'Land Use, Settlement, and Plant Diversity in Iron Age Northwest France', *The Holocene* 28, no. 4 (1 April 2018): 513–28, https://doi.org/10.1177/0959683617735590; Daniele Colombaroli et al., 'Changes in Biodiversity and Vegetation Composition in the Central Swiss Alps during the Transition from Pristine Forest to First Farming', *Diversity and Distributions* 19, no. 2 (2013): 157–70, https://doi.org/10.1111/j.1472-4642.2012.00930.x; Björn E. Berglund et al., 'Long-Term Changes in Floristic Diversity in Southern Sweden: Palynological Richness, Vegetation Dynamics and Land-Use', *Vegetation History and Archaeobotany* 17, no. 5 (1 September 2008): 573–83, https://doi.org/10.1007/s00334-007-0094-x.

22. Jennifer J. Crees et al., 'Millennial-Scale Faunal Record Reveals Differential Resilience of European Large Mammals to Human Impacts across the Holocene', *Proceedings of the Royal Society B: Biological Sciences* 283, no. 1827 (30 March 2016): 20152152, https://doi.org/10.1098/rspb.2015.2152.
23. Alison Burns et al., 'Footprint Beds Record Holocene Decline in Large Mammal Diversity on the Irish Sea Coast of Britain', *Nature Ecology & Evolution* 6, no. 10 (October 2022): 1553–63, https://doi.org/10.1038/s41559-022-01856-2.
24. David Smith, Geoff Hill and Harry Kenward, 'The Development of Late-Holocene Farmed Landscapes: Analysis of Insect Assemblages Using a Multi-Period Dataset', *The Holocene* 29, no. 1 (1 January 2019): 45–63, https://doi.org/10.1177/0959683618804645.
25. Ove Eriksson, 'Species Pools in Cultural Landscapes – Niche Construction, Ecological Opportunity and Niche Shifts', *Ecography* 36, no. 4 (2013): 403–13, https://doi.org/10.1111/j.1600-0587.2012.07913.x.
26. Attila Németh et al., 'Holocene Mammal Extinctions in the Carpathian Basin: A Review', *Mammal Review* 47, no. 1 (2017): 38–52, https://doi.org/10.1111/mam.12075.

27. Anthony D. Barnosky, *Dodging Extinction: Power, Food, Money, and the Future of Life on Earth* (University of California Press, 2014): Chapter 4: Power, https://doi.org/10.1525/9780520959095.
28. Both meet Barnosky's definition of 'megafauna', which is an animal that weighs more than 44 kg.
29. Tim S. Doherty et al., 'Invasive Predators and Global Biodiversity Loss', *Proceedings of the National Academy of Sciences* 113, no. 40 (4 October 2016): 11261–65, https://doi.org/10.1073/pnas.1602480113.
30. Thomas Giesecke et al., 'Postglacial Change of the Floristic Diversity Gradient in Europe', *Nature Communications* 10, no. 1 (28 November 2019): 5422, https://doi.org/10.1038/s41467-019-13233-y.
31. David Smith and Harry Kenward, 'Roman Grain Pests in Britain: Implications for Grain Supply and Agricultural Production', *Britannia* 42 (November 2011): 243–62, https://doi.org/10.1017/S0068113X11000031.
32. Carolina Perpiña Castillo et al., 'Modelling Agricultural Land Abandonment in a Fine Spatial Resolution Multi-Level Land-Use Model: An Application for the EU', *Environmental Modelling & Software* 136 (1 February 2021): 104946, https://doi.org/10.1016/j.envsoft.2020.104946.
33. Cristina Cremene et al., 'Alterations of Steppe-Like Grasslands in Eastern Europe: A Threat to Regional Biodiversity Hotspots', *Conservation Biology* 19, no. 5 (2005): 1606–18, https://doi.org/10.1111/j.1523-1739.2005.00084.x.
34. Laetitia M. Navarro and Henrique M. Pereira, 'Rewilding Abandoned Landscapes in Europe', *Ecosystems* 15, no. 6 (1 September 2012): 900–912, https://doi.org/10.1007/s10021-012-9558-7.
35. Josh Donlan, 'Re-Wilding North America', *Nature* 436, no. 7053 (August 2005): 913–14, https://doi.org/10.1038/436913a.
36. C. Josh Donlan and Harry W. Greene, 'NLIMBY: No Lions in My Backyard', *Restoration and History* (Routledge, 2009).
37. Navarro and Pereira, 'Rewilding Abandoned Landscapes in Europe'.

38. Margaret Davies, 'Rhosili Open Field and Related South Wales Field Patterns', *The Agricultural History Review* 4, no. 2 (1956): 80–96.
39. Rhiannon Philp, 'Changing Tides: The Archaeological Context of Sea Level Change in Prehistoric South Wales' (PhD Thesis, Cardiff University, 2018), https://orca.cardiff.ac.uk/id/eprint/118952/.
40. Michael Shrubb, *Birds, Scythes and Combines: A History of Birds and Agricultural Change* (Cambridge University Press, 2003): 47.
41. Colin R. Tubbs, 'A Vision for Rural Europe', *British Wildlife* 09 no. 2 (1997): 79–85.
42. Colin R. Tubbs, 'Comment – Wilderness or Cultural Landscapes: Conflicting Conservation Philosophies?', *British Wildlife* 07 no. 5 (1996): 290–296.
43. Rob J. F. Burton and Mark Riley, 'Traditional Ecological Knowledge from the Internet? The Case of Hay Meadows in Europe', *Land Use Policy* 70 (1 January 2018): 334–46, https://doi.org/10.1016/j.landusepol.2017.10.014.
44. These rules vary by country. National governments decide exactly how such agri-environment schemes will be operated.
45. Anna Dahlström, Ana-Maria Iuga and Tommy Lennartsson, 'Managing Biodiversity Rich Hay Meadows in the EU: A Comparison of Swedish and Romanian Grasslands', *Environmental Conservation* 40, no. 2 (2013): 194–205.
46. Dániel Babai and Zsolt Molnár, 'Small-Scale Traditional Management of Highly Species-Rich Grasslands in the Carpathians', *Agriculture, Ecosystems & Environment*, Biodiversity of Palaearctic grasslands: processes, patterns and conservation, 182 (1 January 2014): 123–30, https://doi.org/10.1016/j.agee.2013.08.018; Ivaşcu et al., 'The Traditional Perceptions of Hay and Hay-Meadow Management in a Historical Village from Maramureş County, Romania'.
47. 'Number of Scottish Corncrakes on the Rise Thanks to Partnership Recovery Efforts', RSPB, accessed 22 January 2024, https://www.rspb.org.uk/media-centre/scottish-corncrakes.

48. Dániel Babai, Béla Jánó and Zsolt Molnár, 'In the Trap of Interacting Indirect and Direct Drivers: The Disintegration of Extensive, Traditional Grassland Management in Central and Eastern Europe', *Ecology and Society* 26, no. 4 (22 October 2021), https://doi.org/10.5751/ES-12679-260406.
49. Martin Konvicka et al., 'How Too Much Care Kills Species: Grassland Reserves, Agri-Environmental Schemes and Extinction of Colias Myrmidone (Lepidoptera: Pieridae) from Its Former Stronghold', *Journal of Insect Conservation* 12, no. 5 (1 October 2008): 519–25, https://doi.org/10.1007/s10841-007-9092-7.
50. John Akeroyd, 'Transylvania: Nature and Tradition in Transition.' *Bulletin of the Transilvania University of Brasov. Series IV: Philology and Cultural Studies* (February 2, 2023): 9–24. https://doi.org/10.31926/but.pcs.2022.64.15.3.1.

FIVE: *Our Place*

1. Kirsi Sonck-Rautio, 'Adaptation and Cultural Sustainability of the Winter-Seining Community in the Southwest Finland Archipelago', in *Cultural Sustainability and the Nature-Culture Interface* (Routledge, 2018).
2. Sonck-Rautio, 'Adaptation and cultural sustainability of the winter-seining community in the Southwest Finland Archipelago'.
3. Tero Mustonen and the Members of the Kesälahti Fish Base, 'Weather Change Observations of the Puruvesi Winter Seiners 1996-2015' (Snowchange, July 2015), http://www.snowchange.org/pages/wp-content/uploads/2015/07/Snowchange-Discussion-Paper-7.pdf.
4. Laura Stark, *The Magical Self: Body, Society and the Supernatural in Early Modern Rural Finland* (Academia Scientiarum Fennica, 2006): 258.
5. Sonja Hukantaival, 'The Materiality of Finnish Folk Magic: Objects in the Collections of the National Museum of Finland', *Material Religion* 14, no. 2 (3 April 2018): 183–98, https://doi.org/10.1080/17432200.2018.1443893; see also: Vesa Piludu, 'Bear Hunt Rituals in Finland and Karelia: Beliefs, Songs, Incantations and Magic

Rites', *Men and Bears: Morphology of the Wild* (Accademia University Press, 2020).
6. Vesa-Pekka Herva and Timo Ylimaunu, 'Folk Beliefs, Special Deposits, and Engagement with the Environment in Early Modern Northern Finland', *Journal of Anthropological Archaeology* 28, no. 2 (1 June 2009): 234–43, https://doi.org/10.1016/j.jaa.2009.02.001.
7. Timo Myllyntaus and Timo Mattila, 'Decline or Increase? The Standing Timber Stock in Finland, 1800–1997', *Ecological Economics* 41, no. 2 (1 May 2002): 271–88, https://doi.org/10.1016/S0921-8009(02)00034-4.
8. Olli-Pekka Tikkanen et al., 'Habitat Suitability Models of Saproxylic Red-Listed Boreal Forest Species in Long-Term Matrix Management: Cost-Effective Measures for Multi-Species Conservation', *Biological Conservation* 140, no. 3 (1 December 2007): 359–72, https://doi.org/10.1016/j.biocon.2007.08.020.
9. Ilkka Hanski, 'Extinction Debt and Species Credit in Boreal Forests: Modelling the Consequences of Different Approaches to Biodiversity Conservation', *Annales Zoologici Fennici* 37, no. 4 (2000): 271–80.
10. Tracey Nakamura, 'Peatlands and Associated Boreal Forests of Finland Under Restoration', *NOAA Arctic* (blog), 18 October 2023, https://arctic.noaa.gov/report-card/report-card-2023/peatlands-and-associated-boreal-forests-of-finland-under-restoration/.
11. Sari Holopainen and Aleksi Lehikoinen, 'Role of Forest Ditching and Agriculture on Water Quality: Connecting the Long-Term Physico-Chemical Subsurface State of Lakes with Landscape and Habitat Structure Information', *Science of The Total Environment* 806 (1 February 2022): 151477, https://doi.org/10.1016/j.scitotenv.2021.151477.
12. Tero Mustonen, 'Oral Histories as a Baseline of Landscape Restoration – Co-Management and Watershed Knowledge in Jukajoki River', *Fennia – International Journal of Geography* 191, no. 2 (11 December 2013): 76–91.
13. Tero Mustonen et al., 'Traditional Knowledge in Special Fisheries: The Case of Puruvesi Vendace and Seining', *Reviews in Fish Biology*

and Fisheries 33, no. 3 (1 September 2023): 649–67, https://doi.org/10.1007/s11160-022-09728-5.

14. Aby L. Sène, 'Western Nonprofits Are Trampling Over Africans' Rights and Land', *Foreign Policy* (blog), 22 January 2024, https://foreignpolicy.com/2022/07/01/western-nonprofits-african-rights-land/.
15. Victoria Tauli-Corpuz, Janis Alcorn and Augusta Molnar, 'Cornered by Protected Areas: Replacing "Fortress" Conservation with Rights-Based Approaches Helps Bring Justice for Indigenous Peoples and Local Communities, Reduces Conflict, and Enables Cost-Effective Conservation and Climate Action' (Rights and Resources Initiative, 25 June 2018), https://doi.org/10.53892/EXQC6889.
16. 'Agricultural Land Values in Scotland Show Strong Growth in 2021', Strutt & Parker – Rural Hub, accessed 22 January 2024, https://rural.struttandparker.com/article/farmland-values-in-scotland-surge-due-to-rising-competition/.
17. 'Rewilding Boosts Jobs', Rewilding Britain, accessed 22 January 2024, https://www.rewildingbritain.org.uk/press-hub/rewilding-boosts-jobs-and-volunteering-opportunities-study-shows.
18. Rebecca Bliege Bird and Dale Nimmo, 'Restore the Lost Ecological Functions of People', *Nature Ecology & Evolution* 2, no. 7 (July 2018): 1050–52, https://doi.org/10.1038/s41559-018-0576-5.
19. Unto Salo, 'Suomen esihistoriallinen menneisyys ja sen arkeologinen kuva lappalaiskulttuurin tarjoamien analogioiden valossa', *Sananjalka* 20, no. 1 (1 January 1978): 5–16, https://doi.org/10.30673/sja.86417.
20. 'Erämaa–Finnish', Environment & Society Portal, 16 June 2020, https://www.environmentandsociety.org/exhibitions/wilderness-babel/eramaa-finnish.
21. Tero Mustonen, 'Endemic Time-Spaces of Finland: From Wilderness Lands to "Vacant Production Spaces"', *Fennia – International Journal of Geography* 195, no. 1 (20 June 2017): 5–24, https://doi.org/10.11143/fennia.58971.
22. 'Publication of an application pursuant to Article 50(2)(a) of Regulation (EU) No 1151/2012 of the European Parliament and of

the Council on quality schemes for agricultural products and foodstuffs', *Official Journal of the European Union* 2013/C 140/10, EUR-Lex – 52013XC0518(04) – EN – EUR-Lex (europa.eu)

23. 'Winter Seine Fishing in Lake Puruvesi – Elävä Perintö – Wiki', accessed 22 January 2024, https://wiki.aineetonkulttuuriperinto.fi/wiki/Winter_seine_fishing_in_Lake_Puruvesi.
24. Naomi Sykes, 'Woods and the Wild', *The Oxford Handbook of Anglo-Saxon Archaeology* (OUP, 2011) 327–345.
25. Michael Allan Monk, 'The Plant Economy and Agriculture of the Anglo-Saxons in Southern Britain: With Particular Reference to the "mart" Settlements at Southampton and Winchester' (PhD, University of Southampton, 1977), https://eprints.soton.ac.uk/458491/.
26. Jacqueline Fay, 'The Farmacy: Wild and Cultivated Plants in Early Medieval England', *ISLE: Interdisciplinary Studies in Literature and Environment* 28, no. 1 (1 March 2021): 186–206, https://doi.org/10.1093/isle/isz085.
27. Sykes, 'Woods and the Wild'.
28. Nicholas A. Robinson, 'The Charter of the Forest: Evolving Human Rights in Nature', in *The Impact of Environmental Law* (Edward Elgar Publishing, 2020), 54–74, https://china.elgaronline.com/edcollchap/edcoll/9781839106927/9781839106927.00010.xml.
29. The rights to graze animals, collect wood and cut turf.
30. Helen Sanderson and Hew D. V. Prendergast, 'Commercial Uses of Wild and Traditionally Managed Plants in England and Scotland' (Royal Botanic Gardens, Kew, 2002).
31. Marko Lovrić et al., 'Non-Wood Forest Products in Europe – A Quantitative Overview', *Forest Policy and Economics* 116 (1 July 2020): 102175, https://doi.org/10.1016/j.forpol.2020.102175.
32. Valentina Pavlovna Wasson and Robert Gordon Wasson, *Mushrooms, Russia, and History, Volume 1* (Pantheon, 1957).
33. Christopher Smith, 'The Population of Late Upper Palaeolithic and Mesolithic Britain', *Proceedings of the Prehistoric Society* 58, no. 1 (January 1992): 37–40, https://doi.org/10.1017/S0079497X00004084.

34. Lovrić et al., 'Non-wood forest products in Europe – A quantitative overview'.
35. Lovrić et al., 'Non-wood forest products in Europe – A quantitative overview'.
36. Steven Morris, '"It's Trendy": Wild Garlic Foragers Leave Bad Taste in Mouth of Cornish Residents', *The Guardian*, 24 March 2022, sec. Life and style, https://www.theguardian.com/lifeandstyle/2022/mar/24/its-trendy-wild-garlic-foragers-leave-bad-taste-in-mouth-of-cornish-residents.
37. Robin McKie, 'Free-for-All by Wild Mushroom Pickers Puts Woodland Habitats at Risk', *The Observer*, 3 September 2016, sec. Science, https://www.theguardian.com/science/2016/sep/04/new-forest-bans-fungi-pickers.
38. Marla Emery, Suzanne Martin and Alison Dyke, 'Wild harvests from Scottish woodlands'. (Forestry Commission, Edinburgh, 2006).
39. Oliver Rackham, *The History of the Countryside* (Hachette UK, 2020): 91.

SIX: *Where the Wild Things Were*

1. Rene L. A. Catala, 'Report on the Gilbert Islands: Some Aspects of Human Ecology, Gilbert Islands', 1957, http://repository.si.edu/xmlui/handle/10088/4903.
2. Paul V. Nichols, 'Sharks', *FFA Report 92/66* (Pacific Islands Forum Fisheries Agency, 1992).
3. Nick Perry, 'Pacific Nation of Kiribati Establishes Large Shark Sanctuary', accessed 22 January 2024, https://phys.org/news/2016-11-pacific-nation-kiribati-large-shark.html.
4. Julie Adams and Alison Clark, *Fighting Fibres: Kiribati Armour and Museum Collections* (Sidestone Press, 2018): 93.
5. Joshua Drew, Christopher Philipp, and Mark W. Westneat, 'Shark Tooth Weapons from the 19th Century Reflect Shifting Baselines in Central Pacific Predator Assemblies', *PLOS ONE* 8, no. 4 (3 April 2013): e59855, https://doi.org/10.1371/journal.pone.0059855.
6. Jenny Rowland, *Early Welsh Saga Poetry: A Study and Edition of the Englynion* (D.S. Brewer, 1990): 486.

7. Lee Raye, 'Sea Eagles (Haliaeetus Albicilla) in "Canu Heledd" (the Singing of Heledd)', *Natural History* (blog), 27 April 2014, https://historyandnature.wordpress.com/2014/04/27/eaglesincanuheledd/.
8. D.W. Yalden, 'The older history of the White-tailed Eagle in Britain', *British Birds* 100 no. 8 (2007): 471–480
9. Dewi Lewis, 'Eagles in Wales: A Review of Ornithological Literature, County Avifaunas 1889–2000 and Welsh Place Names', 2021.
10. Roger Lovegrove, Iolo Williams and Graham Williams, *Birds in Wales*, 1st ed. (T & AD Poyser, 2010).
11. Sophie-lee Williams et al., 'An Evidence-Based Assessment of the Past Distribution of Golden and White-Tailed Eagles across Wales', *Conservation Science and Practice* 2, no. 8 (2020): e240, https://doi.org/10.1111/csp2.240.
12. Richard J. Evans, Lorcán O'Toole and D. Philip Whitfield, 'The History of Eagles in Britain and Ireland: An Ecological Review of Placename and Documentary Evidence from the Last 1500 Years', *Bird Study* 59, no. 3 (1 August 2012): 335–49, https://doi.org/10.1080/00063657.2012.683388.
13. Hittell himself was a man with an unorthodox past. He was kicked out of university for taking part in the 'snowball rebellion' of 1848, where students barricaded the main entrance and pelted any passing officials in protest at the banning of secret societies on campus. He failed upwards, going on to study at Yale, where he participated in a ritual burial of Euclid's *Elements*, interring the book in a coffin that he built with other seniors. A tutor was sent undercover to monitor the prank. The students, upon discovering the plant, assaulted the man and chased him back to the dormitory. See: Jon T. Coleman, 'The Shoemaker's Circus: Grizzly Adams and Nineteenth-Century Animal Entertainment', *Environmental History* 20, no. 4 (2015): 593–618.
14. Richard B. Lanman et al., 'The historical range of beaver in the Sierra Nevada: a review of the evidence.' *California Fish and Game* 98, no. 2 (2012): 65–80.

15. Charles D. James and Richard B. Lanman, 'Novel physical evidence that beaver historically were native to the Sierra Nevada', *California Fish and Game* 98, no. 2 (2012): 129–132.
16. Lewis, 'Eagles in Wales: a review'.
17. Jonathan Q. Richmond et al., 'Impacts of a Non-Indigenous Ecosystem Engineer, the American Beaver' (Castor Canadensis), in a Biodiversity Hotspot', *Frontiers in Conservation Science* 2 (2021), https://www.frontiersin.org/articles/10.3389/fcosc.2021.752400.
18. Jennifer Price, *Flight Maps: Adventures With Nature In Modern America* (Basic Books, 2000).
19. Lee Raye, 'Early Modern Attitudes to the Ravens and Red Kites of London', *The London Journal* 46, no. 3 (2 September 2021): 268–83, https://doi.org/10.1080/03058034.2020.1857549; Matt Brown, 'London's Red Kites: How Far Have They Spread?', Londonist, accessed 22 January 2024, https://londonist.com/london/maps/red-kites-london.
20. Daniel Pauly, 'Anecdotes and the Shifting Baseline Syndrome of Fisheries', *Trends in Ecology & Evolution* 10, no. 10 (1995): 430.
21. Eric W. Sanderson, 'A Full and Authentic Reckoning of Species' Ranges for Conservation: Response to Akçakaya et al. 2018', *Conservation Biology* 33, no. 5 (2019): 1208–10, https://doi.org/10.1111/cobi.13399.
22. Cormack Gates (IUCN SSC Bison Rla), Dolly Jørgensen and Keith Aune (Wildlife Conservation Society), 'IUCN Red List of Threatened Species: Bison Bison', *IUCN Red List of Threatened Species* (1 September 2016), https://www.iucnredlist.org/en.
23. Tomasz Samojlik, *Conservation and Hunting: Białowieża Forest in the Time of Kings* (Mammal Research Institute, 2005).
24. 'Assessment of conservation of European bison in Ukraine in view of future conservation actions for the species' (WWF Poland and WWF Ukraine 2019): 11, https://www.wwf.pl/sites/default/files/2019-08/Europan%20bison%20Ukraine%20FINAL.pdf.
25. Samojlik, *Conservation and Hunting*.
26. Wojciech Dajczak et al., 'Should Hunting as a Cultural Heritage Be Protected?', *International Journal for the Semiotics of Law – Revue*

Internationale de Sémiotique Juridique 34, no. 3 (1 July 2021): 803–38, https://doi.org/10.1007/s11196-020-09763-0.

27. Zdzisław Pucek (ed.) et al., *European Bison. Status Survey and Conservation Action Plan* (IUCN 2004), https://www.wwf.pl/sites/default/files/2019-09/Action%20Plan_Pucek%20et%20al..pdf.
28. Pucek et al., *European Bison*.
29. Rafał Kowalczyk et al., 'Movements of European Bison (Bison Bonasus) beyond the Białowieża Forest (NE Poland): Range Expansion or Partial Migrations?', *Acta Theriologica* 58, no. 4 (1 October 2013): 391–401, https://doi.org/10.1007/s13364-013-0136-y.
30. Emilia Hofman-Kamińska and Rafał Kowalczyk, 'Farm Crops Depredation by European Bison (Bison Bonasus) in the Vicinity of Forest Habitats in Northeastern Poland', *Environmental Management* 50, no. 4 (1 October 2012): 530–41, https://doi.org/10.1007/s00267-012-9913-7.
31. Boyko Neov et al., 'New Data on the Evolutionary History of the European Bison (Bison Bonasus) Based on Subfossil Remains from Southeastern Europe', *Ecology and Evolution* 11, no. 6 (2021): 2842–48.
32. Astrid Vik Stronen et al., 'Genomic Variability in the Extinct Steppe Bison (Bison Priscus) Compared to the European Bison (Bison Bonasus)', *Mammal Research* 64, no. 1 (1 January 2019): 127–31, https://doi.org/10.1007/s13364-018-0387-8.
33. Hervé Bocherens et al., 'European Bison as a Refugee Species? Evidence from Isotopic Data on Early Holocene Bison and Other Large Herbivores in Northern Europe', *PLOS ONE* 10, no. 2 (11 February 2015): e0115090, https://doi.org/10.1371/journal.pone.0115090.
34. G. I. H. Kerley, R. Kowalczyk and J. P. G. M. Cromsigt, 'Conservation Implications of the Refugee Species Concept and the European Bison: King of the Forest or Refugee in a Marginal Habitat?', *Ecography* 35, no. 6 (1 June 2012): 519–29.
35. Emilia Hofman-Kamińska et al., 'Adapt or Die – Response of Large Herbivores to Environmental Changes in Europe during the

Holocene', *Global Change Biology* 25, no. 9 (2019): 2915–30, https://doi.org/10.1111/gcb.14733.

36. Jacek Radwan et al., 'An Evaluation of Two Potential Risk Factors, MHC Diversity and Host Density, for Infection by an Invasive Nematode Ashworthius Sidemi in Endangered European Bison (Bison Bonasus)', *Biological Conservation* 143, no. 9 (2010): 2049–53.
37. I. Oquinena Valluerca, 'Analysis of Vegetation Changes Induced by a European Bison Herd in the Kraansvlak Area (2003–2009)' (MS thesis, University of Utrecht, 2011).
38. 'Revitalising Dunes in Kennemerland' (PWN, North Holland, 2014), https://issuu.com/pwn_mc/docs/140868_pwn_brochure_revitalising_du.
39. Nieuwe Wildernis, 'Wisenten in Discussie: Een Dag Lang Discussieren Over Het Conceptplan Het Gaat Goed Als Ze Over Je Praten', *Nieuwe Wildernis* 11, 39–40 (2006): 25–34. https://www.wisenten.nl/sites/default/files/wisenten_in_discussie.pdf.
40. Vitaliy Smagol et al., 'Habitat Characteristics of European Bison (Bison Bonasus) in Ukraine', *European Journal of Wildlife Research* 68, no. 3 (2022): 29.
41. Vitas Marozas et al., 'Distribution and Habitat Selection of Free-Ranging European Bison (Bison Bonasus L.) in a Mosaic Landscape – a Lithuanian Case', *Forests* 10, no. 4 (2019): 345; Miloslav Zikmund et al., 'Habitat Selection of Semi-Free Ranging European Bison: Do Bison Preferred Natural Open Habitats?', *Central European Forestry Journal* 67, no. 1 (2021): 30–34.

SEVEN: *The Laboratory of Time*

1. Christopher J. Williams et al., 'Structure, Biomass, and Productivity of a Late Paleocene Arctic Forest', *Proceedings of the Academy of Natural Sciences of Philadelphia* 158, no. 1 (2009): 107–27; Anne-Marie Tosolini, David Cantrill, 'Fossil Forests under Antarctic Ice', accessed 22 January 2024, https://phys.org/news/2021-03-fossil-forests-antarctic-ice.html; Jason J. Head et al., 'Giant Boid Snake from the Palaeocene Neotropics Reveals Hotter Past Equatorial

Temperatures', *Nature* 457, no. 7230 (2009): 715–17 ;Cécile Mourer-Chauviré and Estelle Bourdon, 'The Gastornis (Aves, Gastornithidae) from the Late Paleocene of Louvois (Marne, France)', *Swiss Journal of Palaeontology* 135, no. 2 (2016): 327–41 ;'Mammal Diversity Exploded Immediately after Dinosaur Extinction', UCL News, 23 December 2015, https://www.ucl.ac.uk/news/2015/dec/mammal-diversity-exploded-immediately-after-dinosaur-extinction.

2. Laura C. Foster et al., 'Surviving Rapid Climate Change in the Deep Sea during the Paleogene Hyperthermals', *Proceedings of the National Academy of Sciences* 110, no. 23 (2013): 9273–76 ; Danielle Fraser and S. Kathleen Lyons, 'Mammal Community Structure through the Paleocene-Eocene Thermal Maximum', *The American Naturalist* 196, no. 3 (2020): 271–90 ; Appy Sluijs et al., 'Subtropical Arctic Ocean Temperatures during the Palaeocene/Eocene Thermal Maximum', *Nature* 441, no. 7093 (2006): 610–13.
3. Vera A. Korasidis and Scott L. Wing, 'Palynofloral Change Through the Paleocene–Eocene Thermal Maximum in the Bighorn Basin, Wyoming', *Paleoceanography and Paleoclimatology* 38, no. 12 (2023): e2023PA004741; 'From Mississippi to Wyoming, Plants Once Danced to Fast-Changing Climate Tune', NSF, accessed 22 January 2024, https://www.nsf.gov/news/news_summ.jsp?cntn_id=104601.
4. Ellen D. Currano, Peter Wilf, Scott L. Wing, Conrad C. Labandeira, Elizabeth C. Lovelock and Dana L. Royer, 'Sharply Increased Insect Herbivory during the Paleocene–Eocene Thermal Maximum', *Proceedings of the National Academy of Sciences* 105, no. 6 (February 12, 2008): 1960–64. https://doi.org/10.1073/pnas.0708646105.
5. Philip D. Gingerich, 'Temporal Scaling of Carbon Emission and Accumulation Rates: Modern Anthropogenic Emissions Compared to Estimates of PETM Onset Accumulation', *Paleoceanography and Paleoclimatology* 34, no. 3 (2019): 329–35.
6. Thomas J. Crowley, 'Are There Any Satisfactory Geologic Analogs for a Future Greenhouse Warming?', *Journal of Climate* 3, no. 11 (1990): 1282–92.

7. Alan M. Haywood et al., 'Are There Pre-Quaternary Geological Analogues for a Future Greenhouse Warming?', *Philosophical Transactions of the Royal Society A: Mathematical, Physical and Engineering Sciences* 369, no. 1938 (13 March 2011): 933–56, https://doi.org/10.1098/rsta.2010.0317.
8. K.J. Willis and G.M. MacDonald, 'Long-Term Ecological Records and Their Relevance to Climate Change Predictions for a Warmer World', *Annual Review of Ecology, Evolution, and Systematics* 42, no. 1 (1 December 2011): 267–87, https://doi.org/10.1146/annurev-ecolsys-102209-144704.
9. Piers M. Forster et al., 'Indicators of Global Climate Change 2022: Annual Update of Large-Scale Indicators of the State of the Climate System and Human Influence', *Earth System Science Data* 15, no. 6 (2023): 2295–2327.
10. K. D. Burke et al., 'Pliocene and Eocene Provide Best Analogs for Near-Future Climates', *Proceedings of the National Academy of Sciences* 115, no. 52 (26 December 2018): 13288–93, https://doi.org/10.1073/pnas.1809600115.
11. Jörg Pross et al., 'Persistent Near-Tropical Warmth on the Antarctic Continent during the Early Eocene Epoch', *Nature* 488, no. 7409 (2012): 73–77.
12. Gary S. Dwyer and Mark A. Chandler, 'Mid-Pliocene Sea Level and Continental Ice Volume Based on Coupled Benthic Mg/Ca Palaeotemperatures and Oxygen Isotopes', *Philosophical Transactions of the Royal Society A: Mathematical, Physical and Engineering Sciences* 367, no. 1886 (2009): 157–68.
13. Bev Banks, 'Warming Could Open U.S. for More "Murder Hornets"', E&E News by POLITICO, 8 May 2020, https://www.eenews.net/articles/warming-could-open-u-s-for-more-murder-hornets/.
14. Jetske G. De Boer and Jeffrey A. Harvey, 'Range-Expansion in Processionary Moths and Biological Control', *Insects* 11, no. 5 (2020): 267.
15. Céline Bellard et al., 'Impacts of Climate Change on the Future of Biodiversity', *Ecology Letters* 15, no. 4 (2012): 365–77.

16. Korasidis and Wing, 'Palynofloral Change Through the Paleocene–Eocene Thermal Maximum in the Bighorn Basin, Wyoming'.
17. K. J. Willis, R.M. Bailey, S.A. Bhagwat, and H.J.B. Birks. 'Biodiversity Baselines, Thresholds and Resilience: Testing Predictions and Assumptions Using Palaeoecological Data', *Trends in Ecology & Evolution* 25, no. 10 (2010): 583–91, https://doi.org/10.1016/j.tree.2010.07.006.
18. David P. G. Bond and Yadong Sun, 'Global Warming and Mass Extinctions Associated With Large Igneous Province Volcanism', in *Geophysical Monograph Series*, ed. Richard E. Ernst, Alexander J. Dickson and Andrey Bekker, 1st ed. (Wiley, 2021), 83–102, https://doi.org/10.1002/9781119507444.ch3.
19. Mark C. Urban, 'Accelerating Extinction Risk from Climate Change', *Science* 348, no. 6234 (May 2015): 571–73, https://doi.org/10.1126/science.aaa4984.
20. Chris D. Thomas et al., 'Extinction Risk from Climate Change', *Nature* 427, no. 6970 (January 2004): 145–48, https://doi.org/10.1038/nature02121.
21. Daniel B. Botkin et al., 'Forecasting the Effects of Global Warming on Biodiversity', *BioScience* 57, no. 3 (1 March 2007): 227–36, https://doi.org/10.1641/B570306.
22. Matheus S. Lima-Ribeiro et al., 'Fossil Record Improves Biodiversity Risk Assessment under Future Climate Change Scenarios', *Diversity and Distributions* 23, no. 8 (2017): 922–33, https://doi.org/10.1111/ddi.12575.
23. Kevin L. Campbell et al., 'Substitutions in Woolly Mammoth Hemoglobin Confer Biochemical Properties Adaptive for Cold Tolerance', *Nature Genetics* 42, no. 6 (June 2010): 536–40, https://doi.org/10.1038/ng.574.
24. 'The "Fuel of Evolution" Is More Abundant than Previously Thought in Wild Animals', EurekAlert!, accessed 22 January 2024, https://www.eurekalert.org/news-releases/953928.
25. Sophie Mowbray et al., 'Eyespot Variation and Field Temperature in the Meadow Brown Butterfly', *Ecology and Evolution* 14, no. 1 (2024): e10842, https://doi.org/10.1002/ece3.10842; Patrik Karell

et al., 'Climate Change Drives Microevolution in a Wild Bird', *Nature Communications* 2, no. 1 (22 February 2011): 208, https://doi.org/10.1038/ncomms1213; Giacomo Rovelli et al., 'The Genetics of Phenotypic Plasticity in Livestock in the Era of Climate Change: A Review', *Italian Journal of Animal Science* 19, no. 1 (14 December 2020): 997–1014, https://doi.org/10.1080/1828051X.2020.1809540; John R. Stewart and Adrian M. Lister, 'Cryptic Northern Refugia and the Origins of the Modern Biota', *Trends in Ecology & Evolution* 16, no. 11 (1 November 2001): 608–13, https://doi.org/10.1016/S0169-5347(01)02338-2.

26. Korasidis and Wing, 'Palynofloral Change Through the Paleocene–Eocene Thermal Maximum in the Bighorn Basin, Wyoming'.
27. Eric A. Riddell et al., 'Plasticity Reveals Hidden Resistance to Extinction under Climate Change in the Global Hotspot of Salamander Diversity', *Science Advances* 4, no. 7 (6 July 2018): eaar5471, https://doi.org/10.1126/sciadv.aar5471.
28. Orly Razgour et al., 'Considering Adaptive Genetic Variation in Climate Change Vulnerability Assessment Reduces Species Range Loss Projections', *Proceedings of the National Academy of Sciences* 116, no. 21 (21 May 2019): 10418–23, https://doi.org/10.1073/pnas.1820663116.
29. Craig Moritz et al., 'Impact of a Century of Climate Change on Small-Mammal Communities in Yosemite National Park, USA', *Science* 322, no. 5899 (10 October 2008): 261–64, https://doi.org/10.1126/science.1163428.
30. Michael Archer et al., 'The Burramys Project: A Conservationist's Reach Should Exceed History's Grasp, or What Is the Fossil Record For?', *Philosophical Transactions of the Royal Society B: Biological Sciences* 374, no. 1788 (4 November 2019): 20190221, https://doi.org/10.1098/rstb.2019.0221.
31. Eleanor C. Saxon-Mills et al., 'Prey Naïveté and the Anti-Predator Responses of a Vulnerable Marsupial Prey to Known and Novel Predators', *Behavioral Ecology and Sociobiology* 72, no. 9 (24 August 2018): 151, https://doi.org/10.1007/s00265-018-2568-5.

32. Hayley Bates, 'Assessing Environmental Correlates of Populations of the Endangered Mountain Pygmy-Possum (Burramys Parvus) in Kosciuszko National Park, New South Wales' (PhD Thesis, UNSW Sydney, 2017), https://doi.org/10.26190/unsworks/3327.
33. G. F. Peterken and M. Game, 'Historical Factors Affecting the Number and Distribution of Vascular Plant Species in the Woodlands of Central Lincolnshire', *Journal of Ecology* 72, no. 1 (1984): 155–82, https://doi.org/10.2307/2260011.

EIGHT: *Redemption*

1. Vallet, Anne, et al. 'Effets à long terme des pratiques agricoles sur les populations d'arthropodes: inventaire du site de Thuilley-aux-Groseilles (54)', *La Mémoire des Fôrets* (2004): 255.
2. Oliver Rackham, *Woodlands* (HarperCollins UK, 2012): 260.
3. Vance T. Holliday and William G. Gartner, 'Methods of Soil P Analysis in Archaeology', *Journal of Archaeological Science* 34, no. 2 (1 February 2007): 301–33, https://doi.org/10.1016/j.jas.2006.05.004; 'The Anglo-Saxon Ship Burial at Sutton Hoo | British Museum', accessed 22 January 2024, https://www.britishmuseum.org/collection/death-and-memory/anglo-saxon-ship-burial-sutton-hoo.
4. Tessa T. Baber 'Early Travellers and the Animal "Mummy Pits" of Egypt: Exploration and Exploitation of the Animal Catacombs in the Age of Early Travel', S. Porcier, S. Ikram, S. Pasquali (éd.), *Creatures of Earth, Water, and Sky: Essays on Animals in Ancient Egypt and Nubia, Leyde* (2019): 67–86.
5. Lander Baeten, Martin Hermy, and Kris Verheyen, 'Environmental Limitation Contributes to the Differential Colonization Capacity of Two Forest Herbs', *Journal of Vegetation Science* 20, no. 2 (2009): 209–23, https://doi.org/10.1111/j.1654-1103.2009.05595.x.
6. E. Dambrine et al., 'Present Forest Biodiversity Patterns in France Related to Former Roman Agriculture', *Ecology* 88, no. 6 (2007): 1430–39, https://doi.org/10.1890/05-1314.
7. Abdala G. Diedhiou et al., 'Response of Ectomycorrhizal Communities to Past Roman Occupation in an Oak Forest', *Soil*

Biology and Biochemistry 41, no. 10 (1 October 2009): 2206–13, https://doi.org/10.1016/j.soilbio.2009.08.005.

8. Jed O. Kaplan, Kristen M. Krumhardt and Niklaus Zimmermann, 'The Prehistoric and Preindustrial Deforestation of Europe', *Quaternary Science Reviews* 28, no. 27 (1 December 2009): 3016–34, https://doi.org/10.1016/j.quascirev.2009.09.028.
9. A. Izdebski et al., 'Palaeoecological Data Indicates Land-Use Changes across Europe Linked to Spatial Heterogeneity in Mortality during the Black Death Pandemic', *Nature Ecology & Evolution* 6, no. 3 (March 2022): 297–306, https://doi.org/10.1038/s41559-021-01652-4.
10. Rebecca Hamilton et al., 'Non-Uniform Tropical Forest Responses to the "Columbian Exchange" in the Neotropics and Asia-Pacific', *Nature Ecology & Evolution* 5, no. 8 (August 2021): 1174–84, https://doi.org/10.1038/s41559-021-01474-4.
11. Krzysztof Stereńczak et al., 'ALS-Based Detection of Past Human Activities in the Białowieża Forest – New Evidence of Unknown Remains of Past Agricultural Systems', *Remote Sensing* 12, no. 16 (January 2020): 2657, https://doi.org/10.3390/rs12162657.
12. Alison M. Derry, Peter G. Kevan and Susan D. M. Rowley, 'Soil Nutrients and Vegetation Characteristics of a Dorset/Thule Site in the Canadian Arctic', *Arctic* 52, no. 2 (1999): 204–13.
13. Marianne S. V. Douglas et al., 'Prehistoric Inuit Whalers Affected Arctic Freshwater Ecosystems', *Proceedings of the National Academy of Sciences* 101, no. 6 (10 February 2004): 1613–17, https://doi.org/10.1073/pnas.0307570100.
14. 'Ponds', *Freshwater Habitats Trust* (blog), accessed 22 January 2024, https://freshwaterhabitats.org.uk/habitats/ponds/.
15. Oospores are the sexual spore produced in some algae.
16. J. Shen-Miller et al., 'Exceptional Seed Longevity and Robust Growth: Ancient Sacred Lotus from China', *American Journal of Botany* 82, no. 11 (1995): 1367–80, https://doi.org/10.2307/2445863.
17. Svetlana Yashina et al., 'Regeneration of Whole Fertile Plants from 30,000-y-Old Fruit Tissue Buried in Siberian Permafrost',

Proceedings of the National Academy of Sciences 109, no. 10 (6 March 2012): 4008–13, https://doi.org/10.1073/pnas.1118386109.

18. Dustin Wolkis et al., 'Germination of Seeds from Herbarium Specimens as a Last Conservation Resort for Resurrecting Extinct or Critically Endangered Hawaiian Plants', *Conservation Science and Practice* 4, no. 1 (January 2022): e576, https://doi.org/10.1111/csp2.576.
19. Emily Alderton et al., 'Buried Alive: Aquatic Plants Survive in "Ghost Ponds" under Agricultural Fields', *Biological Conservation* 212 (August 2017): 105–10, https://doi.org/10.1016/j.biocon.2017.06.004.
20. 'Rare Grass-Poly Rediscovered in Norfolk "Ghost Pond"', UCL News (30 November 2020), https://www.ucl.ac.uk/news/2020/nov/rare-grass-poly-rediscovered-norfolk-ghost-pond.

NINE: *Hallowed Ground*

1. Emergency motion for an injunction pending appeal under circuit rule 27–3, *Apache Stronghold v. United States of America*, No. 21-15295 (United States Court of Appeal Ninth Circuit, 23rd February 2021), https://becketnewsite.s3.amazonaws.com/Apache-Stronghold-Motion-for-Injunction-Pending-Appeal-file-stamped.pdf.
2. 'FINAL Environmental Impact Statement: Resolution Copper Project and Land Exchange', Volume 1 (US Department of Agriculture, January 2021), https://www.resolutionmineeis.us/sites/default/files/feis/resolution-final-eis-vol-1.pdf.
3. T. J. Ferguson, Maren P. Hopkins and Chip Colwell, 'Oak Flat is an Important Cultural Site for Nine Tribes: The Resolution Copper Mine will Impact Hundreds of Tribal Traditional Cultural Properties', *U.S. House of Representatives Natural Resources Committee Subcommittee on Indigenous Peoples of the United States* (Tuesday, April 13, 2021), https://docs.house.gov/meetings/II/II24/20210413/111424/HHRG-117-II24-20210413-SD013.pdf.
4. Kalliopi Stara, Rigas Tsiakiris and Jennifer L.G. Wong, 'The Trees of the Sacred Natural Sites of Zagori, NW Greece', *Landscape*

Research 40, no. 7 (3 October 2015): 884–904, https://doi.org/10.1080/01426397.2014.911266.

5. John Halley, John Healey and Kalliopi Stara, 'Sacred Sites Have a Biodiversity Advantage That Could Help World Conservation', The Conversation, 17 May 2018, http://theconversation.com/sacred-sites-have-a-biodiversity-advantage-that-could-help-world-conservation-95599.
6. 'The Church Forests of Ethiopia', Emergence Magazine, 11 January 2020, https://emergencemagazine.org/feature/the-church-forests-of-ethiopia/.
7. Charles Stewart, *Demons and the Devil: Moral Imagination in Modern Greek Culture* (Princeton University Press, 2016): 165–166.
8. 'Newt-counting delays are a massive drag on the prosperity of this country,' lamented prime minister Boris Johnson in 2020.
9. Dominic Alexander, *Saints and Animals in the Middle Ages* (Boydell & Brewer Ltd, 2008): Chapter 7.
10. Alexander, *Saints and Animals in the Middle Ages*: Chapter 7.
11. Alexander, *Saints and Animals in the Middle Ages*: Chapter 8.
12. 'Eider Duck: Marine Conservation Zones – Hansard – UK Parliament', 22 January 2024, https://hansard.parliament.uk/Commons/2018-02-23/debates/202AF057-0A02-4351-B3B3-4044167F71F8/EiderDuckMarineConservationZones.
13. Antone Minard, 'The Mystery of St Cuthbert's Ducks: An Adventure in Hagiography', *Folklore* 127, no. 3 (2016): 325–43.
14. Angus Macvicar, *Golf In My Gallowses* (Arrow, 1987): 34.
15. Literary references are often invoked in a similar manner. The chequerboard landscape of Otmoor, a wetland outside Oxford, is often said to have inspired Lewis Carroll's *Through the Looking-Glass* – something that was at the centre of a campaign by Friends of the Earth when the government proposed expanding a motorway through it in the 1980s. But when I attempted to find a source for the tale, I came away empty-handed. 'Long story short, there's no evidence for it,' Franziska Kohlt, editor of the *Lewis Carroll Review*, told me.

16. Huw Pryce, 'A new edition of the *Historia Divae Monacellae*', *The Montgomeryshire Collections* 82 (1994): 23–40.
17. Shooting Gazette, 'Pheasant Shooting at Llechweddygarth, Powys', *ShootingUK* (blog), 11 October 2012, https://www.shootinguk.co.uk/features/pheasant-shooting-at-llechweddygarth-powys-10163/.
18. Jim Perrin, 'Country Diary: No Sanctuary for Hunted Partridge at Melangell's Church | Hunting | The Guardian', accessed 22 January 2024, https://www.theguardian.com/uk-news/2018/jul/14/country-diary-no-sanctuary-hunted-partridge-melangells-church-llangynog-powys.
19. Nicholas Hellen, 'Pilgrims' Fury at Llangynog Valley Shoot Carnage', accessed 22 January 2024, https://www.thetimes.co.uk/article/pilgrims-fury-at-shoot-carnage-90qdkl9xz.
20. Marianne Brooker, 'Pheasant Shooting Is "Massive Waste of Life"', 1 October 2018, https://theecologist.org/2018/oct/01/pheasant-shooting-massive-waste-life-reports-marianne-brooker.
21. Tim M. Blackburn and Kevin J. Gaston, 'Abundance, Biomass and Energy Use of Native and Alien Breeding Birds in Britain', *Biological Invasions* 20, no. 12 (1 December 2018): 3563–73, https://doi.org/10.1007/s10530-018-1795-z.
22. Lucy A. Capstick, Rufus B. Sage and Andrew Hoodless, 'Ground Flora Recovery in Disused Pheasant Pens Is Limited and Affected by Pheasant Release Density', *Biological Conservation* 231 (1 March 2019): 181–88, https://doi.org/10.1016/j.biocon.2018.12.020.
23. R. B. Sage et al., 'The Flora and Structure of Farmland Hedges and Hedgebanks near to Pheasant Release Pens Compared with Other Hedges', *Biological Conservation* 142, no. 7 (1 July 2009): 1362–69, https://doi.org/10.1016/j.biocon.2009.01.034.
24. The petition in question relates to a proposal to build a commercial clay pigeon shooting facility in the valley, which occurred at the same time as the extension of the shoot at Llechweddygarth. The two issues clearly became somewhat conflated at the time, and many of those commenting on the petition also referenced the intensification of the pheasant shoot.

25. Chirstopher Graffius, 'The Facts about Powys Game Shoot', *The Guardian*, 25 July 2018, sec. Letters, https://www.theguardian.com/environment/2018/jul/25/the-facts-about-powys-game-shoot.
26. Pryce, 'A new edition of the *Historia Divae Monacellae*'.
27. W. J. Britnell et al., 'Excavation and Recording at Pennant Melangell Church', *The Montgomeryshire Collections* 82 (1994): 41–102.
28. Jeremy Harte, 'At the Edge: How Old Is That Old Yew?', accessed 22 January 2024, https://www.indigogroup.co.uk/edge/oldyews.htm.
29. Jack Hunter, 'The Folklore of the Tanat Valley: Fairies, Giants and Forgotten Ecological Knowledge', *Newsletter of the Fairy Investigation Society* (Jan 2022): 31–47.
30. 'Clwyd-Powys Archaeological Trust – Projects – Historic Landscapes – The Tanat Valley – Funerary Landscapes', accessed 22 January 2024, https://www.cpat.org.uk/projects/longer/histland/tanat/tnfunera.htm.
31. Jack Hunter, *Manifesting Spirits* (Aeon Books, 2020).
32. For the uninitiated, this is a reference to the character of the Dude in *The Big Lebowski*.
33. Ronald Fuge, Catherine F. Paveley and Matthew T. Holdham, 'Heavy Metal Contamination in the Tanat Valley, North Wales', *Environmental Geochemistry and Health* 11, no. 3 (1 December 1989): 127–35, https://doi.org/10.1007/BF01758662.
34. R. Richards, 'Some Giant Stories of the Upper Tanat Valley', *The Montgomeryshire Collections* 43 (1934): 168.

EPILOGUE: *Old Souls*

1. Leda Cosmides and John Tooby, 'Evolutionary Psychology: A Primer' (1993), https://www.cep.ucsb.edu/evolutionary-psychology-a-primer/.
2. E. O. Wilson, 'Biophilia and the conservation ethic,' in S. R. Kellert and E. O. Wilson (eds.), *The Biophilia Hypothesis*. (Washington, DC: Island Press 1993): 31–41.
3. Wilson, 'Biophilia and the conservation ethic'.
4. G. Stanley Hall, 'A Study of Fears', *The American Journal of Psychology* 8, no. 2 (1897): 147–249, https://doi.org/10.2307/1410940.

5. Eleonora Gullone, 'The Development of Normal Fear: A Century of Research', *Clinical Psychology Review* 20, no. 4 (1 June 2000): 429–51, https://doi.org/10.1016/S0272-7358(99)00034-3.
6. Vanessa LoBue, 'Perceptual biases for threat', in *Psychology of Bias* (Hauppauge, NY: Nova Science Publishers 2012): 37–52.
7. *Small Voices, Big Dreams 2012: A global survey of children's hopes, aspirations and fears* (ChildFund Alliance 2012), https://reliefweb.int/report/world/small-voices-big-dreams-2012.
8. Gullone, 'The development of normal fear: A century of research'.

Acknowledgements

I spoke to many archaeologists, palaeoecologists, conservationists and others in the course of writing this book. Some of them are named within its pages, and I am grateful to all of them for their time and expertise. I am particularly thankful to those who gave hours or days of their time to meet me in person: to Philip and Myrtle Ashmole, for welcoming me into their home and showing me around Carrifran; to Tero and Kaisu Mustonen, to Karoliina Lehtimäki and Lauri Hämäläinen, and to the rest of the Snowchange team, for allowing me a glimpse into their world; to Carl Sayer for showing me a ghost pond; to Mark Hipkin for taking me back to medieval Wales; to Jack Hunter for the many laps of the church.

There are others who provided me with knowledge, sources and references, whom I was not able to quote directly, but whose words had no less of an impact on my research and thinking. And so my thanks also to Jennifer Crees, Nicki Whitehouse, Ralph Fyfe, Jessie Woodbridge, Anne de Vareilles, Ove Eriksson, Petr Kuneš, Ross Barnett, David Hetherington, Martyn Barber,

Jonathan Last, Matthew Hay, Hannah Fluck, Mark Bell, Răzvan Popa, Mick Drury, Jenny Wong, Samuel Hudson, Claudine Gerrard and Elena Pearce.

Thank you to my agent, Tim Bates, and to the team at HarperNorth for giving physical form to my words – and for the patience. I am also grateful to Lee Schofield, Patrick Barkham, Guy Shrubsole and Chris Packham for the kind words in advance of publication, and to Harriet Rix for her botanical wisdom and enthusiasm upon reading drafts of certain chapters. My heartfelt thanks to Coreen Grant, who has worked alongside me at Inkcap Journal almost since the beginning; without her dedication and brilliance, I would never have found the time to write this book. And thank you to the readers of Inkcap Journal, whose support and encouragement have been invaluable over these past few years.

Thank you to my friends and family, who have not questioned my disappearance too much. In particular to Jen, whose friendship has never wavered, and to Emilia, for the moral support and waffles. To my Mum for always giving me the time and encouragement to write, and especially for all the love, and to my Grandma, who is the most generous-hearted person I know. To Jack and Amy for the sibling rivalry: would I have even bothered writing this book if not for you two? And to my Hayling family, upon whom I can always count to be my biggest cheerleaders.

Thank you to Clara, who made writing this book a hundred times harder and my life a million times better. Witnessing your delight at the world is my greatest joy. And, finally, my deepest gratitude to my husband Jack: for the time, the car rides, the route planning, the conversations, the wisdom, the love. I could write a book about how much you have done for me, but let's not go there.

Index

Harper North

Book Credits

HarperNorth would like to thank the following staff and contributors for their involvement in making this book a reality:

Fionnuala Barrett
Samuel Birkett
Peter Borcsok
Ciara Briggs
Katie Buckley
Sarah Burke
Fiona Cooper
Alan Cracknell
Jonathan de Peyer
Anna Derkacz
Tom Dunstan
Kate Elton
Sarah Emsley
Nick Fawcett
Simon Gerratt
Lydia Grainge
Monica Green
Natassa Hadjinicolaou
Megan Jones
Jean-Marie Kelly
Taslima Khatun
Anna Lord
Holly Macdonald
Dan Mogford
Petra Moll
Alice Murphy-Pyle
Adam Murray
Genevieve Pegg
Natasha Photiou
James Ryan
Florence Shepherd
Eleanor Slater
Emma Sullivan
Katrina Troy